Research in Biotechnology

Principles of Experimental Design in Biotechnology
Rock Canyon High School

Volume 4
May 2019

Editors
Shawndra L. Fordham
Susanne M. Petri
Bryan M. Winkelman

COVER PHOTO CREDITS

Middle back: "A fruit fly (*Drosophila melanogaster*) feeding off a banana" by Sanjay Acharya, 2017 (https://commons.wikimedia.org/wiki/File:Drosophila_melanogaster_Proboscis.jpg) CC BY SA 4.0
All other cover photos courtesy of our student researchers

Cover designed by Zoë Zizzo and Sarah Bermingham

ISBN: 9781092727020
Independently Published

ACKNOWLEDGMENTS

The success of the fourth year of Rock Canyon High School's Experimental Design in Biotechnology course has been the result of a combined effort of so many people, from our generous research funders, to our amazing mentors, and everyone in between. First, we would like to give a special thank you to Bryan Winkelman, RCHS Teacher Librarian, for his continued collaboration, innovative ideas, design of the website, and help with publication of our journal and scientific posters. This program would not be possible without the guidance of our scientific expert mentors. We would like to extend a special thank you to Kathryn Scott, a graduate student in the Cell Biology, Stem Cells, and Development program at University of Colorado Anschutz Medical Campus and RCHS Research Partnership Program Coordinator for her efforts to connect our research students with expert mentors to support them with their research. We would also like to thank the following RCHS teachers for volunteering their time to review and give feedback on the students' research proposals, help students perform lab protocols, evaluate chemical safety of their research, and perform statistical analysis of their data: Jeff Seaquist, David Ferguson, Kerry Hinton, Kyler Barker, Daniel Jibson, Nikki Dobos, and Gwendolyn Karaba. We are grateful to the RCHS science department, school administration, and Douglas County School District for their continued support of this course and program.

We would finally like to thank all of the families, friends, donors, and sponsors who contributed to the students' projects. Many of you have asked to remain anonymous, so we will not recognize you by name, but please know how much we appreciate your contributions. For the rest, the students will thank you personally in their individual acknowledgments, but we want to recognize the following donors and sponsors who donated $100 or more to the students' research this year:

DONORS
The Naik Family
Tewell Warren Printing
Jonathan Manske
The DeMarte Family
The Elango Family
The Kozlowski Family
Dr. Christopher Link
The Appel Lab: Kathryn Scott
The Boyle Lab: Dana Dabelesa
The Rytis Prekeris Lab: Emily Duncan
Dr. Gary Elliott, Biocolor Ltd.
Robert Kos - Harvard Bioscience
RCHS Science Department

CONTENTS
May 2019

FOREWORD

Upon arriving at Rock Canyon High School in 2011, I was immediately impressed with the work that was happening with our small Biotechnology program led by instructor Mrs. Shawndra Fordham. As I began to learn more about this program and Shawndra's vision, I found myself in a state of perpetually saying yes and seeking ways to support the growth and opportunity this course provides. It has been an absolute pleasure to watch this program grow, and I have found great reward in watching the amazing learning opportunities our students get to experience.

This program has now grown beyond a part-time position to two full-time instructors with assistance from other experts in our building. These students work with researchers throughout the state and across the country, and the research they have studied and conducted themselves provides them with a state-of-the-art education and experience that is unmatched in the state of Colorado and certainly in the top ten nationally. I am confident that this learning experience will positively impact them beyond their secondary and post-secondary education.

I am extremely proud of both Shawndra and Susanne. They have worked tirelessly to see this vision come to fruition, which I am sure is incredibly rewarding for them. When we get into education, it is to make a difference in the lives of our youth, to make a difference in the world and community around us. I am proud to be a colleague with teachers who has undoubtedly made a large impact in these areas and will clearly continue to do so.

I invite, you, the readers of this journal, to reflect on your own educational experiences. If you are like me, you will find yourself in amazement and awe reading the work of our students. I am truly honored to be the principal at a high school with such an amazing opportunity for students.

Sincerely,

Andy Abner
Principal
Rock Canyon High School

Featured Research

Efficacy of Triton X-100 as a candidate for whole organ decellularization

C. A. Ewing, O. A. Cesarone, S. E. Mellett & S. L. Fordham
Department of Science, Principles of Experimental Design in Biotechnology, Rock Canyon High School, Highlands Ranch, Colorado

Common methods to treat kidney failure have significant drawbacks. Dialysis requires chronic hospitalization, and viable donor organs are extremely scarce and can be immunologically rejected. Contemporary research has focused on the creation of decellularized kidney scaffolds that may be implanted after cell seeding. These naturally obtained scaffolds serve as a promising source of organs for transplantation, as they are compatible with the recipient's immune system and maintain a natural vasculature and protein composition suitable for cell differentiation and function. We investigated the efficacy of Triton X-100 perfusion in removing cellular material from porcine kidneys while leaving an intact extracellular matrix (ECM) and conserving glycosaminoglycans (GAG). Further, we compared data from Triton X-100 trials to kidneys treated with another common solvent, sodium dodecyl sulfate (SDS). We hypothesized that Triton X-100 would be the ideal solvent, relative to SDS, in whole organ decellularization on the bases of cell removal and GAG preservation. Kidney samples were subject to both spectrophotometric GAG testing and histology with 4′,6-diamidino-2-phenylindole (DAPI). Data collected from GAG assays showed that Triton X-100 conserved 48.56% of GAGs, while PBS conserved 22.07% of GAGs relative to untreated kidneys. Based on histological analysis and visible changes in kidney coloration, both Triton X-100 and SDS were successful in removing cellular material from kidneys. These findings lead us to believe that while both Triton X-100 and SDS are suitable for whole organ decellularization, the perfusion of any solution will cause extensive damage to the extracellular matrix and GAGs specifically.

With over 114,860 people in need of a life-saving organ transplant and approximately 5,000 donors annually, there is an obvious need for solutions to the worldwide organ shortage crisis. The transplant waiting list continues to grow, and many patients die waiting for an organ they will never receive. In 2006 alone, more than 6% of patients on the waiting list died. The large number of patients on the waiting list and the unavailability of the organs has led to an increase in cost for medical treatments such as dialysis (Abouna, 2008). The kidney is in especially high demand, as kidney transplants have accounted for over 58% of all transplants in the United States since 1988 (United Network for Organ Sharing, 2018). Not only are donor kidneys scarce, but transplantation presents a great risk in immune system incompatibility, particularly chronic rejection of the allographic tissue. The rejection stems from what is known as "indirect presentation" of alloantigens from antigen-presenting cells to T-cells, activating an attack against the foreign tissue (**Fig. 1**). There are pharmaceutical treatments available to suppress the immune system, but a great majority of them are intrinsically toxic and generate an increased susceptibility to infection (Ruiz *et al.,* 2013). When considering the large discrepancy between patients needing organ transplants and the availability of donor organs, scientists have investigated the potential of xenotransplantation, especially from pigs to humans. While this would solve the organ shortage, xenotransplantation has

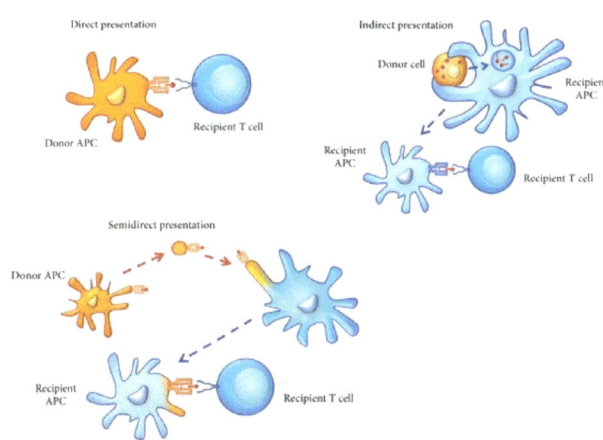

Figure 1: The three types of alloantigen presentation to the recipient immune system, with the most harmful and untreatable (indirect presentation) at the top-right. Reprinted from "Transplant Tolerance: New Insights and Strategies for Long-Term Allograft Acceptance," by P. Ruiz, 2013, *Clinical and Developmental Immunology, 2013*(210506). Copyright 2013 by P. Ruiz. CC BY 3.0

not been possible to date due to severe host-graft rejection, the presence of harmful viruses in the porcine tissue, and both physiological and molecular incompatibilities between species (Iwase & Kobayashi, 2015). One solution to this issue is being pursued through gene editing technology such as CRISPR. The aim of CRISPR efforts have primarily been to deactivate the harmful viruses held within the porcine tissue, such as porcine endogenous retroviruses (PERVs). By using the Cas9 editing enzyme, the pig genome can be edited to prevent replication of all copies of PERV, stopping transmission into humans and thus reducing the risk of disease contraction. However, this method has not been widely employed because of its inability to prevent harmful immune responses in graft recipients (Yang *et al.,* 2015). Other methods to mitigate immune rejection, such as regulatory T-cell therapy, have similarly failed to generate long-term acceptance of donor tissues (Monguio-Tortajada, Lauzurica-Valdemoros, & Borras, 2014). A much more attractive long-term solution to the problem has risen in the field of tissue engineering. The engineering of tissue to create skin grafts is already a successful and common practice in medicine today. More complex organs such as the kidney, heart, and lungs also have great potential to be engineered in the laboratory setting. By applying a wide range of solvents, organs and tissues can be decellularized, leaving an extracellular matrix (ECM) that may be restored to function by being reseeded with a patient's own stem cells. This technique is most commonly used on organs in high demand for transplantation such as the heart, liver, kidney, and lungs, as the complete removal of native cellular material may allow for the elimination of immune rejection (Badylak, Taylor, & Uygun, 2011).

Obtaining an acellular scaffold is the first step in being able to engineer an entire functional organ using stem cells. The next step, seeding stem cells onto the scaffold, requires that the obtained ECM demonstrate certain characteristics representative of that organ *in vivo*. In order for stem cells to correctly differentiate and repopulate the scaffold, there must be suitably high levels of structural ECM proteins, collagens in particular, and glycosaminoglycans (GAGs) remaining after the decellularization process. Collagens act as fibrous proteins that have numerous adhesive and structural roles in the ECM (**Fig. 2**), while GAGs are negatively charged

polysaccharides that form porous hydrogels, contributing to the mechanical structure of the ECM and allowing it to withstand compressive forces. The preservation of these compounds is vital in ensuring that seeded stem cells obtain nutrients, signal one another, and are exposed to growth factors to induce appropriate cellular differentiation (Destefani, Sirtoli, & Nogueira, 2017). Though there is a general consensus on the characteristics of an ideal final organ scaffold, numerous studies have suggested differing methodologies in whole organ decellularization. Various enzymatic and chemical methods have been researched extensively with regard to not only their efficacy in removal of cellular material, but also in how they affect the ECM and the compounds that are left behind. Commonly used solvents include Triton X-100, trypsin, and SDS, as they have been shown to effectively decellularize various organs without adversely affecting the resulting biologic scaffold (Crapo, Gilbert, & Badylak, 2011). In recent years, testing of these solvents has shifted from smaller, rodent-based organs to porcine organs. Porcine organs are comparable in size to those of humans, and the scaffolds of pig-based organs have been shown to adhere to human cells and encourage more growth and differentiation than scaffolds generated from dogs or other primates (Destefani *et al.*, 2017). Among all of this testing, researchers have struggled to come to an agreement on the most advantageous solvent for each type of organ.

The kidney has been one of the most troublesome organs in terms of detailing an effective decellularization method. Numerous studies have come to different conclusions about how exactly to go about removing cellular material from the kidney, which, as previously stated, is the single most demanded organ in transplantation. One study suggested that Triton X-100 solutions were the most suitable solvents for whole porcine kidney decellularization. This decision was made with consideration to the structural and biomechanical integrity of the resulting scaffold, taking GAGs into account (Choi *et al.*, 2015). Yet another study found that using a 0.5% SDS solution was best suited for whole porcine kidney decellularization on the basis of conservation of critical ECM components such as collagens and GAGs (Sullivan *et al.*, 2012). Thus, there is a need for a more a thorough understanding of the effects that each of these solvents have on GAGs and collagen during decellularization. Studies from BYU's Tissue Engineering Lab largely suggest that GAGs are more susceptible to destruction in the decellularization procedure than proteins like collagen, and are therefore crucial indicators of solvent performance (Poornejad *et al.*, 2016). In our research, we tested the ability of Triton X-100 to effectively remove cellular material from porcine kidneys while preserving GAGs. Further, we compared the performance of Triton X-100 to SDS in GAG preservation. Considering the fact that SDS-treated tissues provide inaccurate GAG readings in many spectrophotometry-based tests, these comparisons were made using chiefly qualitative data regarding cell removal (Reing *et al.*, 2010). We hypothesized that Triton X-100 would be the ideal solvent, relative to SDS, in whole organ decellularization on the bases of cell removal and GAG preservation.

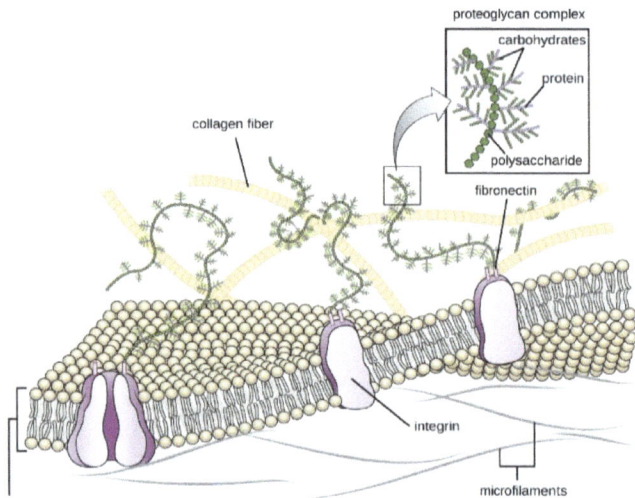

Figure 2: A detailed diagram of the ECM outlines both collagen fibers and the relationship between GAGs and the proteoglycan complex. Reprinted from CNX OpenStax, 2016, Retrieved, April 6, 2019, from https://cnx.org/contents/5CvTdmJL@4.4:DSUpINfV@4/Unique-Charac teristics-of-Eukaryotic-Cells. Copyright 2016 by CNX OpenStax Microbiology. CC BY4.0

Research regarding the decellularization of organs has been primarily limited to professional or university settings, and the field is still in its infancy. The first successful internal organ decellularization was performed in 2008, where an entire heart was exposed to perfusion of detergent solvents (Ott *et al.*, 2008). Our investigation served as a proof-of-concept study that organ decellularization (**Fig. 3**) is a practice that may be carried out in secondary institutions as well.

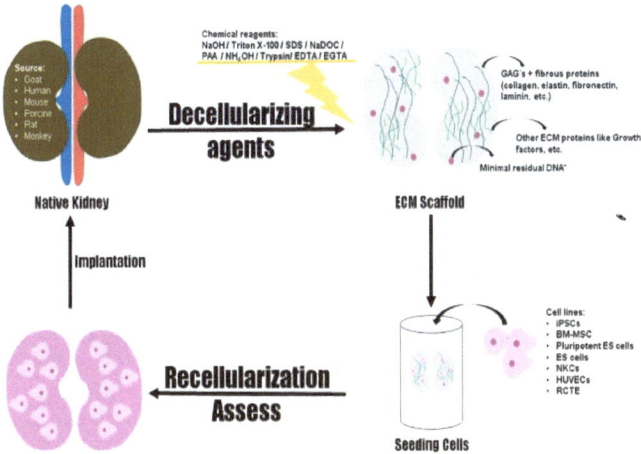

Figure 3: The process by which organs may be decellularized and subsequently restored to function. Reprinted from "Advances in the Knowledge about Kidney Decellularization and Repopulation," A. C. Destefani, 2017, *Frontiers in Bioengineering and Biotechnology, 5*, 34. Copyright 2017 by A. Destefani. CC BY 4.0

To perform this investigation, we obtained porcine kidneys, with the blood vessels intact, from Oliver's Meat Market located in Denver, Colorado, and Innovative Foods LLC located in Evans, Colorado. All methods that were used to decellularize the kidneys were based on protocols found in published literature. We perfused the porcine kidneys with 0.5% solutions of SDS and Triton X-100, and analyzed the efficacy of each solution in removing cellular material, without adversely affecting the conservation of GAGs, and leaving an intact organ scaffold. To perform perfusion, a silicone tube was cannulated through the renal artery, and a peristaltic pump provided pressure to supply a continual flow of solvent through the renal vasculature. The presence of native cell material, DNA specifically, was analyzed before and after perfusion using DAPI stain and viewed under the EVOS FL inverted microscope. Additionally, we tested for GAG conservation and distribution using the Blyscan GAG Assay Kit (Biocolor Ltd, Newtownabbey, UK), with higher GAG conservation indicating more success in the preservation of the ECM structural integrity.

METHODS
Kidneys were obtained from Oliver's Meat Market in Denver, Colorado and Innovative Foods LLC in Evans, Colorado and were then split into two control and two experimental groups (n=5). The kidneys were perfused with either Triton X-100 and SDS, or the control solvent PBS. Decellularized kidneys were then subject to GAG testing and

DAPI staining and compared to control and untreated kidneys.

Kidney Retrieval
Porcine kidneys were purchased from local butchers shortly after slaughter. To ensure efficient cannulation, all kidneys were removed with roughly 4" of the renal artery still intact. Excess fat surrounding the cortex was stripped immediately following kidney retrieval (**Pic. 1**). All kidneys were subsequently stored at -20°C until decellularization could be performed. Kidneys were assigned to one of four treatment groups: native (untreated), PBS, SDS, and Triton X-100 (n=5 for each solution).

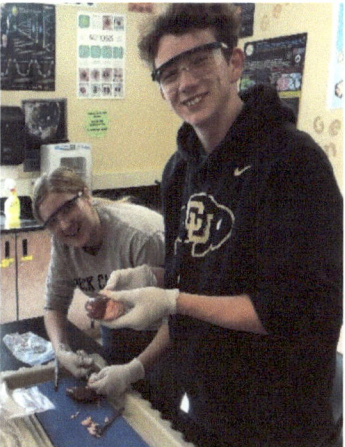

Picture 1: Ewing and Mellett are separating and removing fat from the kidneys to locate the renal artery.

Perfusion Apparatus
Our perfusion apparatus was powered by a Hewlett Packard E3630A Triple Output DC Power Supply set to 6.1V. The output was directed into two peristaltic liquid pumps connected to white silicone tubing (Adafruit Industries, New York City). The tubing was inserted into micropipette tips which were then cannulated

Picture 2: The pumping apparatus supported decellularization of two kidneys at a time.

into the renal artery to provide antegrade solvent flow. Kidneys were set atop grated baking sheets, and 5 gallon plastic buckets stored both used and unused solvent (**Pic. 2**).

Decellularization Protocol
Kidneys were thawed overnight at 4°C before treatment. Immediately before decellularization, 10 kU of heparin was added to 1 liter of 1X PBS solution and perfused via the renal artery at 30 mL/min for 2 minutes to prevent blood coagulation. Solutions of either 0.5% SDS, 0.5% Triton X-100, or 1X PBS were then perfused through each kidney at a constant flow rate of 30 mL/min for 10 hours. The waste from SDS-treated kidneys was disposed of in the sink and the waste from Triton X-100-treated kidneys was stored in waste containers until it was removed by Douglas County Waste Management. After each trial, the containers and the silicone tubing were rinsed with DI water. Treated kidneys

were stored at -20°C until ECM characterization tests could be performed.

Histological Analysis

To test for the presence of residual nuclei, randomly selected kidneys from each group (n=1) were subject to DAPI staining. Thawed tissue samples of roughly 1cm² and negligible width taken from the cortex were immersed in a 1:1 acetone/methanol solution for 15 minutes to achieve cell fixation. Samples were subsequently washed with 1X PBS. Afterwards, fixed cells were immersed in 0.5% Triton X-100 for 5 minutes to break open the cell membranes. Samples were again washed with 1X PBS. Samples were then completely covered with DAPI stain that had been diluted to a 200 nM concentration with 1X PBS and incubated in darkness for 5 minutes. The stain was then removed and samples were washed twice with 1X PBS. An EVOS FL microscope was used to image the samples with a LPlanFL PH2 20x/0.40 ∞/1.2 objective with a DAPI Light Cube.

Glycosaminoglycan Quantification

In order to quantify glyco-saminoglycan preservation levels, we used a Blyscan GAG Assay Kit (Biocolor Ltd, Newtownabbey, UK). Renal cortex samples from all treatment groups (n=5 for each) were cut with a scalpel to a weight of 20-50 mg. A papain extraction reagent was prepared by adding 400 mg sodium acetate, 200 mg EDTA disodium salt and 40 mg cysteine HCL to 50 mL of 0.2M sodium phosphate buffer. Samples were then incubated in the papain extraction solution at 65°C for 3 hours. After 3 hours, the solution was clear and the GAG-dye complex was precipitated using a shaking incubator. The Blyscan dye reagent was added and the solution was centrifuged at 12,000 rpm for 10 minutes. The Blyscan dye dissociation rea-gent was added to the complex and absor-bance of the resulting solution was mea-sured at 655 nm using an iMark micro-plate reader (BioRad Laboratories, Hercules, CA).

RESULTS

Cell Removal

Decellularization of porcine kidneys (n=5) was com-pleted with three separate methods. The control group of kidneys treated with 1X PBS for 10 hours appeared pale brown, and microscopic imaging of the DAPI stained tissue resulted in the highest levels of residual DNA and the lowest cell removal levels when compared to the two treatment groups but less than a native kidney receiving no perfusion treatment at all. Triton X-100 and SDS perfusion resulted in high amounts of cell removal, supported by significant whitening of the kidneys and low incidence of residual DNA remaining as evidenced by low fluorescence after DAPI staining (**Pic. 3**). This data was for descriptive qualitative purposes and was not quantified in terms of relative fluorescence.

GAG Preservation

GAG preservation levels were measured in samples from the Triton X-100 and PBS groups and were compared to samples from native kidneys. Kidneys treated with Triton X-100 preserved 48.6% of GAGs while kidneys treated with PBS only preserved 22.1% of GAGs, as compared to the readings from untreated kidneys. Sulfated GAG concentrations of native kidneys averaged 0.124 μg/mg of wet tissue, PBS-treated kidneys averaged 0.0274 μg/mg, and Triton X-100-treated kidneys averaged 0.0602 μg/mg (**Graph 1**).

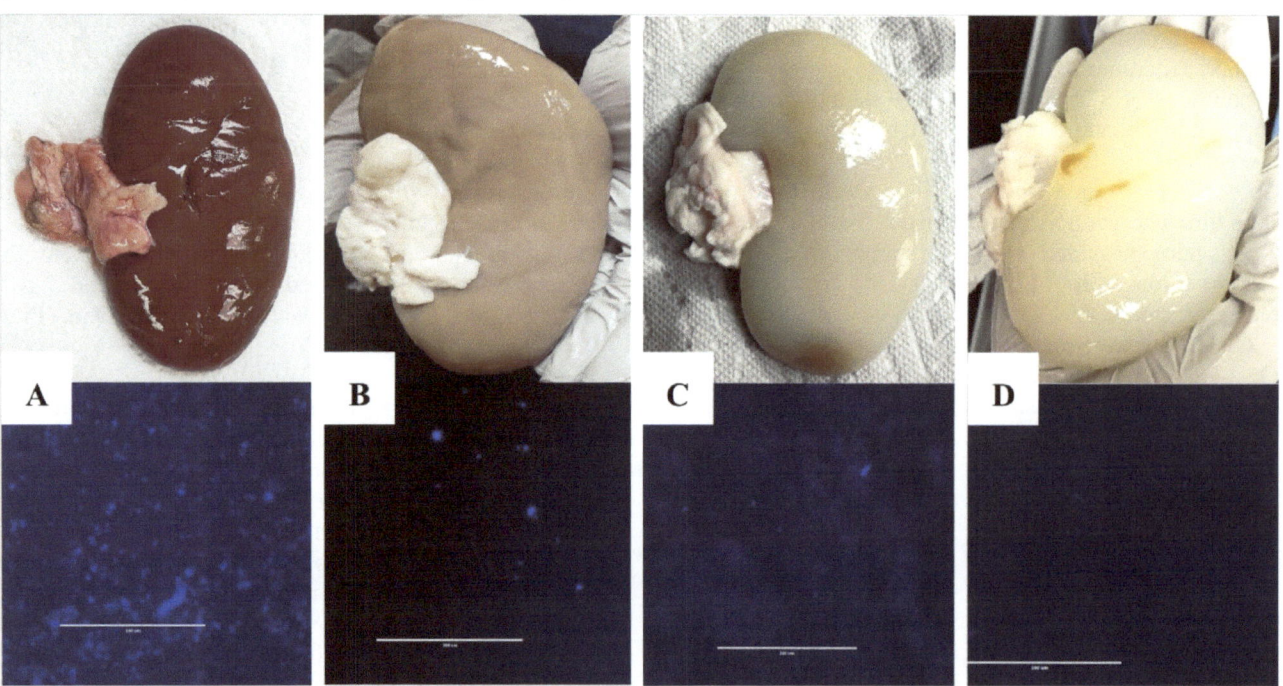

Picture 3: In this image, the top pictures show the progression from a native kidney to a fully decellularized kidney with treatment in different solvents. The bottom images are sections of each kidney stained and imaged with DAPI to show the residual nuclei. Kidney **A** was the control with no treatment; kidney **B** was treated with PBS; kidney **C** was treated with Triton X-100; kidney **D** was treated with SDS.

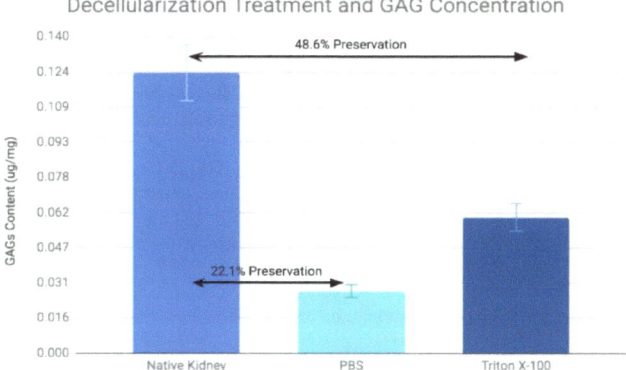

Graph 1: This graph comparing the effects of each solvent on GAG levels demonstrates that Triton X-100 preserved 26.5% more GAGs than PBS treated kidneys.

An unpaired t-test was used to determine statistical significance between treatment groups. When comparing Triton X-100-treated and PBS-treated kidneys with native kidneys, the differences in GAG contents were statistically significant with all p-values <0.001. GAG content could not be measured in SDS-treated samples as the dimethyl-methynlene blue (DMMB) used in the Blyscan assay kit to bind to sulfated GAGs also binds to SDS, resulting in inaccurate GAG concentration readings for those samples.

DISCUSSION
With an average of 20 patients on the organ transplant waiting list dying every day (American Transplant Foundation, 2019), increasing accessibility to donor organs and lessening the incidence of immune rejection in patients is a pressing matter. Whole organ decellularization, the process by which a chemical solvent is perfused through an organ to isolate the ECM, is an emerging solution to this problem. The resulting biologic organ scaffold can then be reseeded with the recipient's own cells, decreasing the likelihood of immune rejection and eliminating the need for tissue-type matching. In an investigation of SDS and Triton X-100 as candidates to decellularize whole porcine kidneys, we hypothesized that Triton X-100 would be the ideal solvent on the bases of cell removal and GAG preservation.

Both Triton X-100 and SDS appeared to remove a majority of cellular material from the kidneys. The drastic color change of kidneys treated with these solvents is also a clear indicator of a substantial loss of cellular material and exposure of the ECM. Resultant scaffolds from SDS-treated kidneys were a slightly brighter, more vibrant white color than those treated with Triton X-100. This suggested that SDS more thoroughly stripped the kidneys of cellular material. Kidneys treated with PBS, conversely, did not experience any significant cell loss, and color changes in the cortex were much less pronounced. The pale brown coloration of the resultant scaffolds can be attributed to the flushing of blood, rather than a loss of renal cells, from the kidney. DAPI staining, which illuminates AT-rich regions of dsDNA, corroborated observations made with the naked eye. DAPI revealed less residual DNA in SDS-treated tissues than in all other treatment groups, though significant loss of DNA

was also observed in Triton X-100-treated tissues. Regardless of the solvent, however, most kidneys retained small pockets of unaffected tissue, suggesting inadequate flow of solvent to those areas. This lack of flow can be attributed to blockage within the vasculature due to either thrombosis or collapse of the arterial tissue itself. Thus, our protocols must be improved upon to ensure as close to complete cell removal as possible, as even minute quantities of dead cells can trigger harmful immune system responses (Nagata, Hanayama, & Kawane, 2010). The consensus among professional researchers is that organ scaffolds suitable for *in vitro* cell culture and subsequent implantation should not have any nuclear material revealed in a DAPI stain (Crapo *et al.*, 2011). By this standard, kidneys treated with SDS appear most suitable for implantation, but a more holistic and quantitative examination of the tissue is required to determine exact levels of residual nuclear material.

With regard to preservation of GAGs, Triton X-100 was more effective than PBS-treated and SDS-treated kidneys were not evaluated with this assay. On average, kidneys treated with Triton X-100 preserved 48.56% of GAGs relative to native kidneys, while the control PBS-treated kidneys only preserved 22.07%. With nearly half of GAGs being preserved, kidneys treated with Triton X-100 would be sufficiently healthy to support the growth of human renal cortical tubular epithelium (RCTE) cells as evidenced by a prior study (Poornejad *et al.*, 2016). In that investigation, researchers at BYU successfully grew RCTE cells on kidneys that had experienced only 30.5% preservation of GAGs. However, scaffolds with GAG preservations closer to 100% are much more conducive to proper cell growth and differentiation (Bosman & Stamenkovic, 2003). Interestingly, the damage caused to the kidneys by PBS suggests that the perfusion of any liquid, even those of neutral pH with an osmolarity similar to most biological materials, has the potential to damage GAGs within the ECM by means of mechanical stress. As previously noted, SDS-treated samples were not tested with the GAG Assay due to the limitations of the DMMB dye.

The perfusion of Triton X-100 for 10 hours with a preemptive heparinized PBS wash sufficiently preserves GAGs as to allow for *in vitro* cell culture. Further, observations made with the naked eye and DAPI staining suggest that Triton X-100 is less effective in cell removal than SDS. Due to the interference of residual SDS in GAG testing, however, any comparative advantage between the two solvents in GAG preservation could not be concluded from our investigation. Additionally, the extensive damage to the ECM caused by the perfusion of a 1X PBS solution suggests that the retrograde flow of any liquid, even if it is nonreactive, through the vasculature of an organ is inherently harmful. Thus, the ideal organ decellularization protocol likely involves a combination of different detergents and a minimal perfusion time.

This experiment was limited by the inability to accurately compare GAG preservation of the SDS-treated samples to Triton X-100- treated samples because SDS binds to the DMMB dye used in the Blyscan GAG assay kit. As a result of time constraints within this experiment, the scaffold could not be reseeded and therefore the viability of this method for restoration of function to the kidney could not be determined. Inadequate solvent flow to small portions of the tissue was likely a result of collapsed arterial tissue or thrombosis within the vasculature. Thus, kidney scaffolds generated through our research would not be suitable for subsequent cell culture and implantation due to the presence of untreated, cellular regions of tissue. Additionally, components other than the GAGs that make up the ECM, such as collagen and proteoglycans could not be tested due to budget and time constraints and therefore a full characterization of ECM preservation could not be performed.

To better understand the efficacy of each solvent, a more thorough investigation of the resultant ECM must be conducted. Various components of the ECM such as growth factors, collagens, and dsDNA must be analyzed to draw more accurate conclusions regarding the success of a decellularization treatment. Further, *in vitro* cell culture of different cell lines upon the scaffold must be observed, as this is the most critical step in restoring function to the organ. Other future steps include analyzing the various effects of solvent exposure time and/or the utilization of multiple solvents on the tissue, along with investigating the performance of these engineered organs in mammals.

ACKNOWLEDGMENTS

We thank Dr. Alonzo Cook, director of the Tissue Engineering Lab at Brigham Young University, Provo with a PhD in Chemical Engineering, and Dr. Emily Beck, a postdoctoral fellow in Bioengineering at the University of Colorado's Anschutz Medical Campus, for their invaluable expertise and advice throughout our investigation. We would like to provide a special thank you to Biotechnology and technical writing teacher Mrs. Susanne Petri for helping to mentor us throughout the entirety of the research process, and providing support and assistance with writing, editing and problem solving and to Mr. Bryan Winkelman, Teacher Librarian, for providing publishing and editing assistance, writing resources, managing the class website and research donations, and support with the design of our scientific poster. We greatly appreciate Dr. Gary Elliott, who graciously donated our Blyscan Glycosaminoglycan Assay Kit from Biocolor Ltd., Mr. Mike Howren and Mr. Matt Bernstein, who kindly gave us a discount on items purchased from Thermo Fisher Scientific, Mr. Dallas Mohler, who lent us the Hewlett Packard power supply, and Mr. Nayan Naik and the Rock Canyon Science Department for funding our research and continually supporting the Rock Canyon High School Biotechnology Program. We would also like to thank the following Rock Canyon High School science teachers: Mr. Daniel Jibson for guidance in designing our perfusion apparatus, Mr. David Ferguson for both donating chemical reagents and managing chemical safety, and Ms. Gwen Karaba for advising our statistical analysis. Last, we would like to thank Rock Canyon High School and Douglas County School District for supporting the Biotechnology Program and providing the laboratory space and equipment needed for successful completion of this research.

REFERENCES

Abouna, G. M. (2008). Organ Shortage Crisis: Problems and Possible Solutions. *Transplantation Proceedings, 40*(1), 34-38. doi:10.101 6/j.transproceed.2007.11.067

American Transplant Foundation. (2019). Facts and Myths about Transplant Retrieved from https://www.americantransplant foundation.org/about-transplant/facts-and- myths/

Badylak, S. F., Taylor, D., & Uygun, K. (2011). Whole-Organ Tissue Engineering: Decellularization and Recellularization of Three-Dimensional Matrix Scaffolds. *Annual Review of Biomedical Engineering, 13*(1), 27-53. doi: 10.1146/annurev-bioeng-071910-124743

Bosman, F. T., & Stamenkovic, I. (2003). Functional structure and composition of the extracellular matrix. *The Journal of Pathology.* 200, 423–428. doi:10.1002/path.1437

Choi, S. H., Chun, S. Y., Chae, S. Y., Kim, J. R., Oh, S. H., Chung, S. K., . . . Kwon, T. G. (2015). Development of a porcine renal extracellular matrix scaffold as a platform for kidney regeneration. *Journal of Biomedical Materials Research Part A, 103*(4), 1391-403. doi: 10. 1002/jbm.a.35274

Crapo, P. M., Gilbert, T. W., & Badylak, S. F. (2011). An overview of tissue and whole organ decellularization processes. *Biomaterials, 32*(12), 3233-3243. doi: 10.1016/j.biomaterials.2011.01.057

Destefani, A. C., Sirtoli, G. M., & Nogueira, B. V. (2017). Advances in the Knowledge about Kidney Decellularization and Repopulation. *Frontiers in Bioengineering and Biotechnology, 5*, 34. doi:10.3389/ fbioe.2017.00034

Iwase, H., & Kobayashi, T. (2015) Current status of pig kidney xenotransplantation. *International Journal of Surgery, 23*(B), 229-233. doi: 10.1016/j.ijsu.2015.07.721

Monguió-Tortajada, M., Lauzurica-Valdemoros, R., & Borràs, F. E. (2014). Tolerance in organ transplantation: from conventional immunosuppression to extracellular vesicles. *Frontiers in Immunology, 2014*(5), 416 doi: 10.3389/fimmu.2014.00416

Nagata S., Hanayama R., & Kawane K. (2010). Autoimmunity and the clearance of dead cells. *Cell*, 140(5), 619–630.10.1016/j.cell.2010. 02.014

Ott, H. C., Matthiesen, T. S., Goh, S.-K., Black, L. D., Kren, S. M., Netoff, T. I., Taylor, D. A. (2008). Perfusion-decellularized matrix: using nature's platform to engineer a bioartificial heart. *Nat. Med.* 14, 213–221. doi:10.1038/nm1684

Poornejad, N., Montahan, N., Salehi, A., Scott, D., Fronk, C., Roeder, B., . . . Cook, A. (2016). Efficient decellularization of whole porcine kidneys improves reseeded cell behavior. *Biomedical Materials, 11*(2), 025003. doi: 10.1088/1748-6041/11/2/025003

Reing, J., Brown, B., Daly, K., Freund, J., Gilbert, T., Hsiong, S., . . . Badylak, S. (2010). The Effects of Processing Methods upon Mechanical and Biologic Properties of Porcine Dermal Extracellular Matrix Scaffolds. *Biomaterials, 31*(33), 8626–8633. doi:10.1016/j. biomaterials.2010.07.083

Ruiz, P., Maldonado, P., Hidalgo, Y., Gleisner, A., Sauma, D., Silva, C., ... Bono, M. R. (2013). Transplant Tolerance: New Insights and Strategies for Long-Term Allograft Acceptance. *Clinical and Developmental Immunology, 2013*(210506), doi: 10.1155/2013/ 210506

Sullivan, D. C., Mirmalek-Sani, S. H., Deegan, D. B., Baptista, P. M., Aboushwareb, T., Atala, A., & Yoo, J. J., (2012). Decellularization methods of porcine kidneys for whole organ engineering using a high-throughput system. *Biomaterials, 33*, 7756–7764. doi:10.1016/ j.biomaterials.2012.07.023

United Network for Organ Sharing. (2018). Matching organs, Saving Lives. Retrieved from https://unos.org/

Yang, L., Guell, M., Niu, D., George, H., Lesha, E., Grishin, D., . . . Church, G. (2015). Genome-wide inactivation of porcine endogenous retroviruses (PERVs). *Science*, 350(6264), 1101-1104. doi:10.1126/ science.aad119

ABOUT THE AUTHORS

Pictured: Here are Ewing, Cesarone, and Mellett at the beginning of the 2018/2019 school year. Not pictured are mentors Dr. Alonzo Cook and Dr. Emily Beck.

Throughout this year, Ewing, Cesarone, and Mellett have learned not only how to conduct their own independent research, but also how to effectively communicate in both professional and personal settings. They have seen the many problems that can arise in experimentation and have learned to quickly overcome them and persevere in order to successfully carry out their research. When faced with difficulties, the group was driven towards the finish line by sheer curiosity. In the future, they hope to use the invaluable skills that they have gained to inspire those around them and make a difference in their communities.

Ewing is incredibly appreciative for the Rock Canyon Biotechnology Program and is excited to perform more research throughout undergraduate studies at the University of Colorado Boulder. Experimental Design in Biotechnology provided the perfect platform to develop critical laboratory and technical writing skills. Inspired by his grandfather Dr. Fred R. Plecha, Ewing's long-term goal is to attend medical school and apply his findings in a clinical setting. Ewing currently volunteers in the ER at Sky Ridge Medical Center.

Cesarone is extremely grateful for her time in the Rock Canyon High School Biotechnology Program. This opportunity has not only confirmed her love for Biomedical Engineering, which she will be majoring in at Baylor University in the fall, but also allowed her to learn more about herself and grow in her confidence. In the future, Cesarone hopes to enter into the medical field and continue down the path of regenerative medicine.

Mellett has enjoyed being a part of the Rock Canyon High School Biotechnology Program these past two years. This program has helped her grow not only as a scientist, but as a person. Mellett's favorite part of being in the Experimental Design in Biotechnology course has been working with other group members and being able to achieve success. This course has helped her discover her love for research and engineering while pushing her drive for medicine all while learning invaluable lab and communication skills. Next fall, Mellett will be attending the University of Colorado Boulder to study Chemical and Biological Engineering.

Effects of exposure to Myelin Sheath Support on oligodendrocyte migration as an indicator of central nervous system myelination in GFP transgenic *Danio rerio* embryos

M. E. Ashbeck, A. C. DeMarte, & S. M. Petri
Department of Science, Principles of Experimental Design in Biotechnology, Rock Canyon High School, Highlands Ranch, Colorado, USA

Demyelinating diseases, such as Multiple Sclerosis (MS), affect over 2.5 million people worldwide and have no direct cure. Current treatments focus on managing the symptoms while finding therapeutics to stimulate the production of new myelin. These aspects are critical in providing effective treatment for these diseases. In neurodegenerative diseases such as MS, myelin, a lipid-based substance contributing to efficiency of neurological processes, is noticeably lacking in the Central Nervous System (CNS). Planetary Herbals claims that its product, Myelin Sheath Support, is able to "promote healthy myelin sheaths." To investigate this non FDA approved claim, we tested the effect of exposure to Myelin Sheath Support on dorsal oligodendrocyte migration in the neural tube in transgenic *Danio rerio* (zebrafish) embryos with GFP tagged oligodendrocyte progenitor cells. Oligodendrocytes are glial cells that myelinate neurons in the CNS. Embryos were exposed to the supplement at 1 day post fertilization (dpf), and oligodendrocyte migration was recorded and imaged using the EVOS-FL microscope at 3 dpf. A higher number of migrated oligodendrocytes demonstrates greater migration which would indicate more myelin production. We hypothesized that early exposure to the supplement would increase oligodendrocyte migration, due to the neuroprotective properties of the ingredients. After analyzing our data using a two-tailed t-test, our results revealed an insignificant difference of oligodendrocyte migration between the control group and Myelin Sheath Support treatment group. Thus, an increase in myelination due to supplement exposure cannot be confirmed and the supplement's claims are not verified by our research.

Neurodegenerative diseases continue to be a pressing issue in medical and biological communities. Demyelination and damage to myelin sheaths are a result of diseases such as MS and Guillain Barre Syndrome, autoimmune diseases that attack myelin in the neurons (Shivane & Chakrabarty, 2007). Symptoms of demyelinating diseases vary but can include numbness or weak-ness, pain and electric shock sensations, loss of vision, and lack of coordination (Mayoclinic, 2017). There is a distinct decrease in oligodendrocyte progenitor cells along with a decrease in myelination in people diagnosed with MS compared to people without these neurodegenerative diseases (Chang, Nishiyama, Peterson, & Trapp, 2000). Treatments that encourage remyelination and repair of myelin sheaths by supporting the production of protective myelin in the CNS can stimulate oligodendrocyte migration, the myelin producing cells, and protect the myelin from degeneration (**Fig. 1**). Identifying drug targets to encourage migration and proliferation of oligodendrocytes will be key to future treatments of demyelinating diseases.

Scientific understanding of demyelination is still in the research stage, and pharmaceutical drugs for demyelinating diseases focus on suppressing inflammatory processes as well as the immune system. Overall, these drugs slow the processes involved in these diseases but do not directly stop

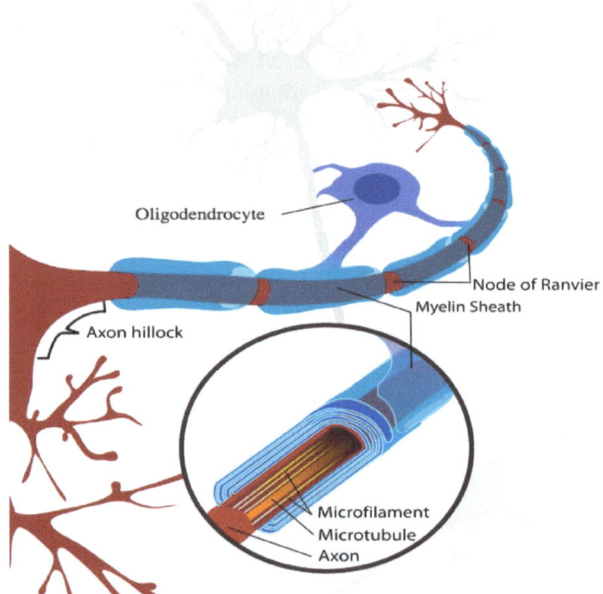

Figure 1: This diagram represents how oligodendrocytes create myelin in the CNS. "Neuron with oligodendrocyte and myelin sheath." Retrieved April 7, 2019, from https://commons.wikimedia.org/wiki/File:Neuron-with_oligodendrocyte_and_myelin_sheath-2.svg. File in public domain.

or repair damage done to the myelin sheath. The focus of new research in this field is to identify drugs that can stimulate oligodendrocyte progenitor cells and protect them from undergoing apoptosis which is a hallmark of these disease processes. Anti-LINGO-1 is a drug currently in clinical research as a possible treatment to slow and repair demyelination. This drug is used to fight MS, as it is an antibody that promotes remyelination. The drug targets protein LINGO, and is able to stop its development, allowing oligodendrocytes to mature. The drug is showing promising results in promoting remyelination in patients with MS, especially when paired with anti-inflammatory drug Tysabri (Almeida, 2016). Anti-LINGO, and other remyelinating drugs being investigated at this time, focus on protecting oligodendrocytes and promoting proliferation of the oligodendrocyte progenitor cells for survival and continue repairing the damaged myelin.

Scientists have completed extensive research on the cellular mechanisms involved in the complex common neurological process of myelination which is most prevailing during an organism's first year of life. Myelin is a white fatty substance, composed of 80% lipids and 20% proteins, that wraps around the axon of the neuron and functions to increase the speed of transmission of electrical impulses and is necessary for proper neurological function (National Multiple Sclerosis Society, 2017). Myelination develops most quickly in the first stages of infancy, in which myelin sheaths are being created around the axons in the CNS, allowing vital brain development (De Graaf Peters & Hadders Algra, 2006). Beyond infancy, myelination still occurs rapidly in the developing years, as oligodendrocytes create myelin sheaths on axons. Myelination directly correlates to properly functioning motor skills, such as walking, as well as cognitive skills. Myelination is mostly complete by the time of early adulthood, but myelin sheaths are still added to regions of grey matter in the brain at any point in life (Chevalier *et al.*, 2015).

Oligodendrocytes, derived from neural stem cells, assist in the production of myelin in the CNS by extending processes from the soma to attach onto axons of surrounding neurons. These processes are then able to produce a series of myelin sheaths around the axons to assist in transmitting neurological signals. Oligodendrocyte

Picture 1: Myelin Sheath Support, by Planetary Herbals, is the supplement we tested in our research.

progenitor cells migrate dorsally to the neural tube during development. In transgenic zebrafish with GFP expressing oligodendrocyte progenitors, scientists can directly observe the migration of oligodendrocytes using fluorescent imaging.

In our research, we tested the effects of a dietary supplement, Myelin Sheath Support by Planetary Herbals, on the dorsal migration of the oligodendrocytes in the neural tube in the CNS of Tg (olig2:EGFP) zebrafish embryos which have GFP expressing oligodendrocyte progenitor cells. Myelin Sheath Support claims to improve the functions of the nervous system by "supporting myelination in the CNS" as well as "promoting a healthier, high functioning nervous system" (Natural Healthy Concepts: Myelin Sheath Support, 2018) (**Pic. 1**). These claims have not been evaluated by the FDA. To measure these effects, we counted the number of oligodendrocytes that migrated to the neural tube using an EVOS-FL microscope with the GFP cube. If this herbal supplement stimulates an increase in oligodendrocyte migration in the CNS, this would indicate an increase in myelination would warrant further research into the effects of this supplement in demyelinating disease models.

Zebrafish are good model organisms for studying neurological development because they are easy to breed, producing up to 200 embryos in a single breeding event. They are also inexpensive and easy to maintain in a lab, and are commonly used as a first step before testing on more complex organisms (Strähle *et al.*, 2012) (**Pic. 2**). We used embryos specifically, because during the first 3 dpf, oligodendrocyte glial cells form and begin to migrate to the neural tube. They are also transparent which allows us to easily view their organs and systems along with their neurological development. By using zebrafish embryos with GFP fluorescent oligodendrocytes, we were able to clearly

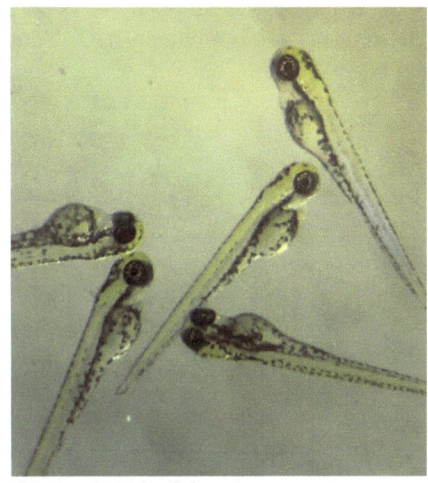

Picture 2: Zebrafish embryos are transparent which allows us to view their organ systems and neurological processes. From *Zebrafish Embryos*, by Kozlowski, 2019. Used with permission.

observe the effects of the supplement on the myelination process. The fluorescence allows us to count oligodendrocyte migration under an EVOS microscope.

Our research tested the effects of the supplement's on oligodendrocyte migration as an indication of myelin production. Myelin Sheath Support contains many herbs including ashwagandha, turmeric rhizome extract, bacopa leaf extract, ginger root, and lion's mane mushroom, all previously shown to have positive neurological properties. Studies of *Withania somnifera* in the CNS have found its extracts to protect against injury of neurons in rats affected with Parkinson's disease (Ahmad *et al.*, 2005); Turmeric

Rhizome Extract improved the regeneration of nerves after leg amputation in mice (Lui, Xu, Li, & Luo, 2016); studies of Bacopa Leaf Extract concluded enhancement in transmission of nerve impulses and aids in repairing neurons, compared to growing cells in a plant tissue culture experiment (Mohan, Jassal, Kumar, & Singh, 2011); while *Zingiber officinale,* commonly known as ginger root, showed support for its use as a potential cognitive and memory enhancer when tested on brain function of middle-aged healthy women (Saenghong *et al.*, 2012). These are just a few examples of the neuroprotective ingredients in Myelin Sheath Support. The intent of this investigation was not to determine the specific ingredient that resulted in any effects observed in the oligodendrocyte migration in the embryos or the mechanism of action of any particular ingredient in Myelin Sheath Support; however, the focus was to determine whether treatment with this supplement holistically impacted oligodendrocyte migration.

To expose the zebrafish embryos to the supplement, we removed the chorion from the embryos at 1 dpf and placed them in egg water infused with the supplement. We tested two different dosages in order to determine if an increase in supplement concentration impacted migration. At 3 dpf, we anesthetized the embryos using tricaine, embedded the embryos in glycerol, and visualized them under the EVOS-FL microscope using the GFP cube. We then counted the number of oligodendrocytes that had begun migration to the neural tube as compared to the control group of embryos not exposed to the supplement. We hypothesized that due to the neuroprotective ingredients in the supplement, the zebrafish embryos would show an increase in oligodendrocyte migration, as myelination is supported.

METHODS

In this research, the effects of the Myelin Sheath Support supplement by Planetary Herbals on oligodendrocyte migration from the yolk extension to the neural tube was measured in transgenic Tg(olig2:EGFP) *Danio rerio*, zebrafish. The oligodendrocytes produced a GFP protein causing them to fluoresce and be visualized using the EVOS-FL microscope using a GPF cube. The embryos were treated with the Myelin Sheath Support supplement through infusion in egg water. The supplement was administered in two dosages to determine the effectiveness at different concentrations. The embryos were exposed to the supplement during our experimental trials, along with a control group with no treatment (egg water only), and a control group with egg water and chemical dimethyl sulfoxide. DMSO was used in both supplement treatments as it assists in the absorption of the supplement by the the embryos. This chemical dissolves both polar and nonpolar compounds and is commonly used as an absorption agent for aquatic life. Although literature has shown amounts of DMSO at or under 1% will not affect mortality or have harmful effects on the embryos, it was included in one of our control groups to account for any difference the DMSO caused in oligodendrocyte migration in our experimental groups (Hallare, Nagel, Köhler, & Triebskorn, 2006).

Embryo Handling and Care

Embryos used in this experiment were donated by University of Colorado Denver Anschutz Medical Campus and stored in petri plates containing egg water in hot water baths set to 28.5°C in a laminar flow hood under ambient lighting conditions (**Pic. 3**). The egg water consisted of 1.5 mL of Instant Ocean Aquarium stock salts and 1 L of distilled water. After imaging the embryos, they were euthanized in a plastic bag filled with egg water and placed at -20°C until

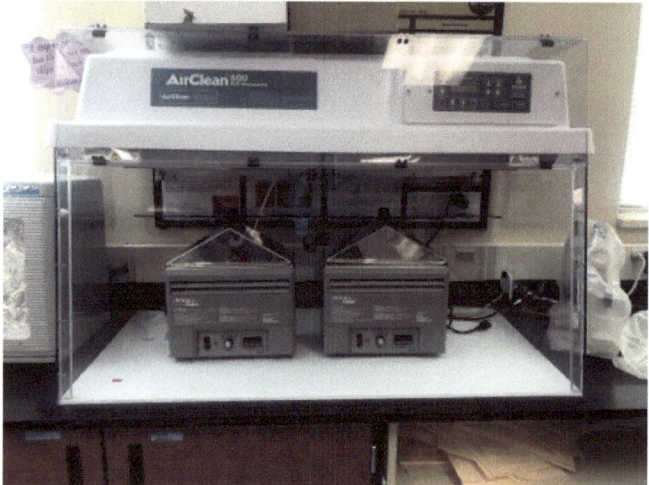

Picture 3: The laminar flow hood with ambient lighting, in which embryos were stored at 28.5° C.

disposed as biological waste, following IACUC guidelines. This method of euthanasia follows the protocols set forth by the National Institute of Health and is considered a humane and ideal way of euthanizing zebrafish embryos (National Institute of Health, 2009).

Experimental Design

In this investigation, the effects of the Myelin Sheath Support on oligodendrocyte migration in Tg(olig2:EGFP) zebrafish embryos was tested. Four trials were performed that included two treatment and two control groups for each trial. One control group consisted of egg water only, while the other control consisted of both egg water and 1% DMSO. The treatment groups included the two different dosages of Myelin Sheath Support determined during pre-trials, 0.079 mg/mL and 0.0395 mg/mL. We determined these two dosages by exposing the embryos to the highest concentration of the supplement, without inducing an increase in mortality. We also tested a lower dosage to understand how the change in dosage relates to oligodendrocyte migration. The first control group tested oligodendrocyte migration of embryos in egg water only while the second control tested the effects of 1% DMSO on oligodendrocyte migration as DMSO was used to increase the uptake of the supplement into the cells of the embryos.

Treatments

At 1 dpf embryos were sorted for fluorescence using the NightSea Attachment on the Leica KL300 LED stereoscope. After this, fluorescing embryos were dechorionated using fine tipped forceps and transferred to an 8 well Corning®

CellBIND® well plate with 15 embryos in each well that contained 4 mL of the respective treatment or control. To prepare the treatments, 256 mg of the Myelin Sheath Support supplement was first dissolved in 16 mL DMSO and then diluted with distilled water to a final concentration of 0.079 mg/mL supplement. This solution was further diluted and DMSO added to bring the second concentration of supplement to 0.035 mg/mL with DMSO. The 1% DMSO control was prepared with 3.96 only control included 4 mL of egg water (**Fig. 2**). 24 hours after treatment, mortality was recorded and dead embryos were removed from treatment wells using a transfer pipette, and disposed of as biohazard waste. At 48 hours post treatment, embryos were removed from their treatment and placed back into plain egg water and prepared for imaging.

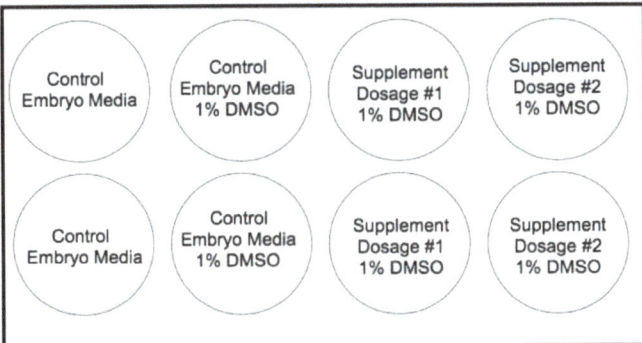

Figure 2: Experimental trials were completed using four treatments (two controls and two experimental). Controls included plain egg water and egg water with 1% DMSO. Experimental trials included both predetermined dosages of Myelin Sheath Support (0.079 mg/mL and 0.035 mg/mL) also with 1% DMSO. Two wells were used for each treatment.

Imaging Oligodendrocytes

At 3 dpf, after 48 hours of exposure to the Myelin Sheath Support supplement, embryos were anesthetized for imaging using 1.5 mL of tricaine per well. The embryos were placed on their side on a coverslip in a drop of 25% glycerol and imaged using the Plan FL PH2 20X/0.40 objective on the EVOS-FL microscope with the GFP light cube (**Pic. 4**).

Picture 4: DeMarte and Ashbeck view an embryo under the EVOS-FL microscope.

Data Analysis

We imaged the region of the neural tube directly above the yolk extension and counted the number of oligodendrocytes that were migrating dorsally (**Pic. 5**). Ten embryos were imaged from each well and their oligodendrocyte migration recorded during each experimental trial. We then determined the average oligodendrocyte migrations for the embryos in each treatment and control group across the four trials and

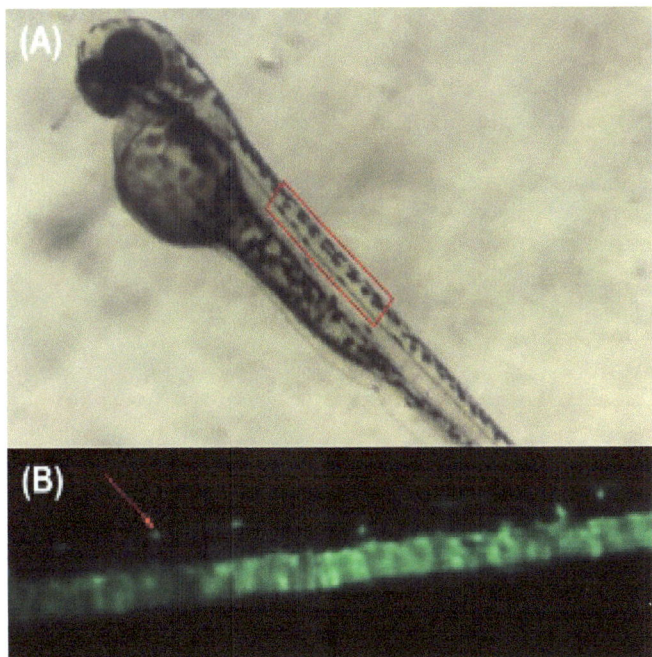

Picture 5: Pictured in image (A) is a zebrafish embryo, with a red box indicating the region of the neural tube directly above the yolk extension in which oligodendrocyte migration was counted. Image (B) shows a zoomed in view of the red box in image (A), which indicated the area above the yolk extension imaged by the EVOS-FL microscope. The green dots seen in the image (B) shows the individual oligodendrocytes that were counted.

the data was analyzed using a two-tailed t-test to determine statistical significance. The egg water only control was compared to the 1% DMSO control and the 1% DMSO control was compared to the two treatment groups. These values were analyzed for statistical significance.

RESULTS

Two dosages of the Myelin Sheath Support supplement were administered to 1 dpf zebrafish embryos to determine its effects on average oligodendrocyte migration (**Graph 1**). The oligodendrocyte migration of the 0.079 mg/mL Myelin Sheath Support treatment group did not change as compared to the 1% DMSO only control with an average of 27.303 and

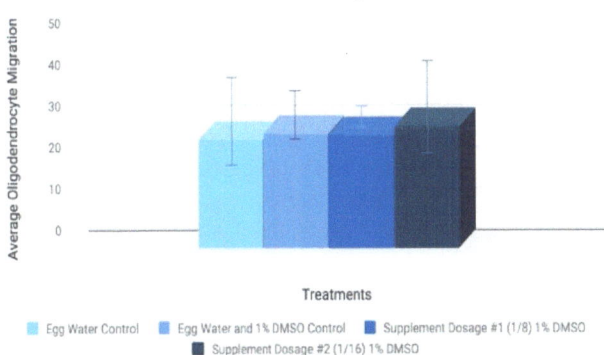

Graph 1: The graph pictured represents the average oligodendrocyte migration of embryos from each treatment group. Because there is little variation between the standard deviation bars, there is not enough difference in the average oligodendrocyte migrations to show significance.

27.657 oligodendrocytes migrating to the neural tube respectively; while the 0.0395 mg/mL Myelin Sheath treatment group had a slight increase in average oligodendrocyte migration compared to the DMSO control with an average of 29.529 migrating oligodendrocytes. These differences were not statistically significant when compared using a two-tailed t-test (p = 0.851 and p = 0.39). In addition, treatment in 1% DMSO only slightly lowered the average oligodendrocyte migration when compared to the egg water only control with an average of 26.145 oligodendrocytes. This difference was also not statistically significant (p = 0.465). Treatment with Myelin Sheath Support supplement in this experiment did not affect the average oligodendrocyte migration in zebrafish embryos at 3 dpf.

DISCUSSION

Demyelinating diseases, such as Multiple Sclerosis (MS), are a pressing issue in today's society. MS alone is a critical condition, affecting 2.5 million people, and with no cure. Demyelinating diseases result in a decrease and destruction of myelin in the nervous system. Myelin is essential as it assists with electrical impulses and the transmission of neurological signals. Oligodendrocytes produce myelin in the CNS, however demyelinating diseases lead to an absence of these cells in this region limiting myelin production. It is essential that we find effective treatments that can directly address the demyelination that occurs with these devastating neurodegenerative diseases.

Myelin Sheath Support (Planetary Herbals) claims its product will assist in promoting healthy myelin sheaths and support myelination in the nervous system. Our research tested the validity of these claims on zebrafish embryos, by counting average oligodendrocytes migrated to the neural tube after 48 hours of supplement exposure. Oligodend-rocytes are the glial cells in the CNS that produce myelin, therefore, an increase in oligodendrocytes migration to the neural tube would indicate an increase in myelination in this region. We hypothesized that the addition of this supplement would increase myelin production and therefore increase oligodendrocyte migration. With 48 hours of treatment in two different doses of the supplement (0.079 mg/mL and 0.0395 mg/mL), no difference was observed in the average oligodendrocyte migration between the treatment and control groups. The 0.079 mg/mL treatment had an average oligodendrocyte migration count of 27.303, and the 0.0395 mg/mL treatment had an average oligodendrocyte count of 29.529. When compared to the average of the DMSO only control of 27.657, this difference was not statistically significant (p = 0.851 and 0.389 respectively). When comparing the egg water control to the 1% DMSO control the average oligodendrocyte migration of 26.145 and 27.657 respectively was also not statistically significant with a p-value of 0.465. Overall, this data did not support our hypothesis that treatment with Myelin Sheath Support supplement would increase oligodendrocyte migration to the neural tube in the CNS of in zebrafish embryos at 3 dpf. We can neither confirm nor deny any change in oligodendrocyte

migration due to Myelin Sheath Support supplement exposure as our p-values state the insignificance of our data.

A source of error in our data could have been due to variation in oligodendrocyte counting. For each embryo, we counted the number of oligodendrocytes directly above the yolk sac extension (**Pic. 6**). However, there were variations in the exact region observed on each embryo. Even a slight movement of the microscope moved the region viewed, causing more or less oligodendrocytes to be counted. There was also ambiguity in a few of the embryos as to what counted as an oligodendrocyte, as some were very faint or blurred, or had just begun migration. These oligodendrocytes were accounted for, but may not have been visible in other embryos, causing variation.

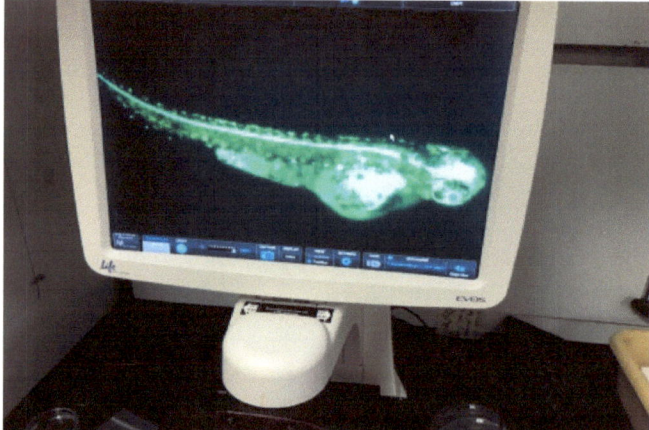

Picture 6: Imaged above is a GFP fluorescing zebrafish embryo. The fluorescence of the embryo allowed us to view oligodendrocyte migration as an indicator of myelination.

Further, the embryos may not have been exposed to the supplement for a sufficient amount of time. Due to time constraints of the IACUC guidelines placed on use of the embryos by CU Anschutz, we were unable to work with them past 3 dpf. This may have led to the supplement not having enough time to begin increasing myelination so repeating the research with longer exposure time could account for this. A longer exposure time to the supplement may have led to changes in oligodendrocyte migration that were significant. The next steps that could be taken with our research would include more experimentation with this supplement to test the claims made by Planetary Herbals. This could include further testing on zebrafish embryos and adult zebrafish. Due to the fact that olig-oligodendrocyte migration in zebrafish embryos is at its maximum during early stages of embryonic development, it would be beneficial to test this supplement on adult zebrafish, as it is known that oligodendrocyte migration begins to slow in adult zebrafish older than 15 months (Münzel, Becker, & Williams, 2014). It would also be useful to test this supplement on diseased models to investigate its myelin repair or protective abilities in a situation where demyelination is known to occur.

Herbal supplements make strong claims on their ability to improve health. It is important to investigate whether these supplements are truly helping one's health or potentially having no effect. The Myelin Sheath Support supplement claims have not yet been investigated by the FDA therefore consumers are unable to know whether the supplement actually improves myelination or is simply an unfounded statement. Ultimately, more research needs to be performed using this supplement on other model organisms other than zebrafish embryos to show the increase in myelination.

ACKNOWLEDGMENTS

We sincerely thank our mentor, Kathryn Scott from Colorado University Anschutz, for providing us with her help on specific skills throughout our research including imaging; and by supplying us with embryos, tricaine, and imaging plates. We would also like to thank our Biotechnology teacher Shawndra Fordham who has been a wonderful instructor and mentor for our research this year as she invested both her time and effort into supporting us with our research. A special thank you to Nayan Naik and the Rock Canyon High School (RCHS) Science Department for providing funding for our project. Additionally, we would like to thank Bryan Winkelman, RCHS Teacher Librarian, who was of great assistance to us this year by helping us with our blog posts throughout the year, finding resources surrounding our research, and with his guidance in our journal publication. We appreciate the following RCHS teachers and students for their support of our research: Chemical Safety Manager Dave Ferguson for reviewing our protocols for safety and assisting us in the process of dissolving our supplement; statistics instructor Gwendolyn Karaba, who helped us analyze our results and for teaching us how to display them correctly in visuals; and former biotechnology students Olivia Landauer, Audrey Gruszczynski, Amanda Wilkins, and Cambri Reisig who provided us with the initial training on the skills we needed to work with zebrafish embryos effectively. Last, we thank both Rock Canyon High School and the Douglas County School District for supporting the Biotechnology Program that provided us the opportunity to perform this research.

REFERENCES

Ahmad, M., Saleem, S., Ansari, A.,Yousuf, F., Hoda, N., & Islam, F. (2005). Neuroprotective effects of *Withania somnifera* on 6-hydroxydopamine induced Parkinsonism in rats. *SAGE Journals, 24*(3), 137-147. doi:10.1191/0960327105ht509oa

Almeida, M.J. (2016, November 10). Anti-LINGO-1 / BIIB033 / Opicinumab for RRMS. *Multiple Sclerosis News Today*. Retrieved from *https://multiplesclerosisnewstoday.com/anti-lingo-1-biib033-opicinumab-for-rrms*

Chang, A., Nishiyama, A., Peterson, J., Prineas, J., & Trapp, B. (2000). NG2-Positive Oligodendrocyte Progenitor Cells in Adult Human Brain and Multiple Sclerosis Lesions. *Journal of Neuroscience. 20*(17) 6404-6412. doi: 10.1523/JNEUROSCI.20-17-06404.2000

Chevalier, N., Kurth, S., Doucette, M. R., Wiseheart, M., Deoni, S. C. L., Dean, D. C., ... LeBourgeois, M. K. (2015). Myelination is associated with processing speed in early childhood: preliminary insights. *PLoS ONE, 10*(10), e0139897. doi: 10.1371/journal.pone.0139897

De Graaf-Peters, V. B. & Hadders-Algra, M. (2006). Ontogeny of the human Central Nervous System. *Early Human Development, 82*(4), 257-266. doi:10.1016 j.earlhumdev.2005.10.013

Hallare, A., Nagel, K., Köhler, H.R., & Triebskorn, R. (2006). Comparative embryotoxicity and proteotoxicity of three carrier solvents to zebrafish (Danio rerio) embryos. *Ecotoxicol Environmental Safety. 63*(3) 378-88. doi: 10.1016/j.ecoenv.2005.07.006

Jansson, L. & Akerman, K. (2014). The role of glutamate and its receptors in the proliferation, migration, and differentiation and survival of neural progenitor cells. *Journal of Neural Transmission. 121*(8) 819-836. doi:10.1007/s00702-014-1174-6

Kozlowski, A. (2019). *Zebrafish Embryos* [JPG]

Lui, G., Xu, K., Li, J. & Luo, Y. (2016). Curcumin upregulates S100 expression and improves regeneration of the sciatic nerve following its complete amputation in mice. *Neural Regeneration Research. 11*(8), 1304-1311. doi: 10.4103/1673-5374.189196

MayoClinic. (2017). Multiple sclerosis. *Diseases and Conditions*. Retrieved from https://www.mayoclinic.org/diseases-condition s/multiple-sclerosis/symptoms- causes

Mohan, N., Jassal, P.S., Kumar, V., & Singh, R.P. (2011). Comparative In vitro and In vivo study of antioxidants and phytochemical content in Bacopa monnieri. *Recent Research in Science and Technology. 3*(9), 78-83. Retrieved from http://scienceflora.org/journals/index.php/rr st/article /view/782/767

Münzel, E. J., Becker, C. G., Becker, T., & Williams, A. (2014). Zebrafish regenerate full thickness optic nerve myelin after demyelination, but this fails with increasing age. *Acta neuropathologica communica teons, 2*, 77. doi:10.1186/s40478-014-0077-y

National Multiple Sclerosis Society. (2017). What is Myelin? Retrieved from https://www.nationalmssociety.org/What-is-MS/ Definition-of-MS/Myelin

Natural Healthy Concepts: Myelin Sheath Support. (2018). Retrieved from https://www.naturalhealthyconcepts.com/myelin-sheath-support-PH180

National Institute of Health (2009). Guidelines for use of zebrafish in the NIH Intramural Research Program. National Institute of Health. Retrieved from https://oacu.oir.nih.gov/sites/default/files/uploads/ara c-guidelines/zebrafish.pdf

Neuroscience News. (2014, April). Researchers find inherited pathway for Schizophrenia. Retrieved from https://neurosciencenews.com /oligo-dendrocytes-schizophrenia-genetics-807/

Saenghong, N., Wattanathorn, J., Muchimapura, S., Tongun, T., Piyavhatkul, N., Banchonglikitkul, C., ... Kajsongkram, T. (2012). Zingiber officinale Improves Cognitive Function of the Middle-Aged Healthy Women. *Evidence-based Complementary and Alternative Medicine*. eCAM.:383062. doi: 10.1155 /2012/383062

Shivane, A., & Chakrabarty, A. (2007). Multiple sclerosis and demyeli-nation. *Current Diagnostic Pathway. 13*(3) 193-202. doi: 10.1016-/j.cdip .2007.04.003

Strahle, U., Scholz, S., Geisler, R., Greiner, P., Hollert, H., Rastegar, S., ... Braunbeck, T. (2012). Zebrafish embryos as an alternative to animal experiments—A commentary on the definition of the onset of protected life stages in animal welfare regulations. *Reproductive Toxicology. 33*(2) 128-132. doi: 10.1016/ J.reprotox.2011.06.121

Wikimedia (2010) Neuron with oligodendrocyte and myelin sheath. Retrieved from https://commons.wikimedia.org/wiki/File:Neuron-_with_oligodendrocyte_and_myelin_sheath-2.svg

ABOUT THE AUTHORS

Pictured (left to right) Pictured above is Scott (mentor), who provided embryos and assistance throughout the year. Ashbeck and DeMarte can be seen to the right.

Ashbeck is a junior at Rock Canyon High School where she completed a year in the Biotechnology Program before joining the Experimental Design in Biotechnology course. Ashbeck has been a part of the Rock Canyon poms program for three years, and hopes to be a captain for her fourth. She has a passion for biology and research, and is extremely thankful for her experience with the biotechnology program this year. She wants to pursue a career in some type of biological field and would not be where she is today without the support from this program.

DeMarte is a senior at Rock Canyon High School. She has taken many upper level science classes such as Human Anatomy and Physiology, Honors Chemistry, and Honors Biology. She completed a year in the Biotechnology program before joining the Experimental Design in Biotechnology course. Next year, she will attend Colorado State University where she will continue exploring her love for science and research and study Biomedical and Biological/Chemical engineering. DeMarte has been on the Rock Canyon poms program for three years, and hopes to continue dancing by joining the CSU dance team. She feels extremely grateful for the opportunity to perform this research.

Throughout their research, this team has learned many valuable lessons such as perseverance and teamwork. The team faced many obstacles, but was able to work past them and succeed. The team also learned the importance of communication, both with one another and their mentor. Aside from the life lessons that were learned throughout the experiment, the team also gained knowledge on how to work with zebrafish embryos in a research setting, as well as how to image, anesthetize, and euthanize zebrafish embryos properly. The team feels they have gained an immense amount of knowledge and discovered their passion for research throughout this process.

Effect of maternal weight on insulin sensitivity in infant umbilical cord mesenchymal stem cells

S. Shubhangi & M. Keleher

Department of Science, Principles of Experimental Design in Biotechnology, Rock Canyon High School, Highlands Ranch, Colorado, USA

Maternal obesity during pregnancy adversely impacts infant health. Research suggests that maternal obesity during pregnancy is a key contributor to increased infant adiposity. It is still not clear if this increase in adiposity is due to greater fat cell number (hyperplasia) or greater adipocyte size (hypertrophy). This study assessed the effect of maternal weight on the growing patterns of adipocytes differentiated from umbilical cord mesenchymal stem cells (MSCs). Mesenchymal stem cells were differentiated into adipocytes and the cells were subsequently stained with BODIPY, Wheat Germ Agglutinin (WGA), and DAPI in order to characterize the cells. The hypothesis was that the adipocytes derived from the umbilical cords of infants from obese mothers (Ob-MSCs) would exhibit increased hypertrophy, not hyperplasia as compared to those derived from normal weight mothers (NW-MSCs). The data showed that the differentiated adipocytes of infants from obese mothers stored more fat, were less in number, and were larger than adipocytes from NW-MSCs, confirming that they exhibited clear patterns of hypertrophy. A t-test conducted on these results demonstrated statistical significance between the Ob-MSCs and NW-MSCs for BODIPY (p = 0.006), WGA (p = 0.030), and DAPI (p = 0.003). Due to the strong link between hypertrophy and the subsequent development of insulin resistance, this study warrants further research into the risk factor of maternal obesity on the later development of type 2 diabetes in infants.

Obesity has been coined as a severe public health issue, as it is directly correlated to metabolic diseases such as type 2 diabetes and cardiovascular disease. According to the Centers for Disease Control and Prevention, in 2017, the prevalence of adulthood obesity in the United States was 39.8%, while the prevalence of childhood obesity was 18.5% (Centers for Disease Control and Prevention, 2018). Despite several efforts to combat this, rates of childhood and adulthood obesity have been rapidly escalating, going from 30.5 to 35.7 percent in between 1999 and 2010 for adults and 13.9 to 18 percent for children. Therefore, it is crucial to find accurate ways to study the mechanisms of this condition.

As mentioned above, obesity has a strong influence on the development of type 2 diabetes. Those who have a BMI greater than 30 kg/m^2, larger amounts of upper body fat, and greater visceral adiposity are highly likely to have abnormal insulin signaling pathways (Spiegelman & Flier, 1996). Type 2 diabetes is linked to changes in the function of adipocytes, which are fat cells that regulate whole body homeostasis and perform several metabolic functions including lipid storage, insulin sensitization, and hormone secretion (Stephens, 2012). In type 2 diabetes, adipocytes secrete lower amounts of adiponectin, a hormone that increases glucose metabolism and insulin sensitization. Furthermore, adipocytes release greater amounts of proinflammatory adipokines, which facilitate a greater release of free fatty acids. Muscle tissue that is exposed to higher amounts of free fatty acids demonstrate impaired insulin stimulated glucose uptake (Rosenberg & Spiegelman, 2006).

Most alarmingly, epidemiological data demonstrates that maternal obesity during pregnancy can significantly affect infant adiposity and obesity rates. By analyzing MSCs that are derived from the umbilical cords of obese mothers, BMI ≥ 30 kg/m^2, and normal weight mothers, BMI of 19-24 kg/m^2, it has been demonstrated that maternal obesity can be directly correlated to accelerated fetal adipogenesis, greater lipid accumulation in adipocytes, and greater infant fat mass. Although it is known that infants born to obese mothers have higher adiposity, it is unclear if this is due to greater fat cell number known as hyperplasia, or greater adipocyte size known as hypertrophy (**Fig. 1**).

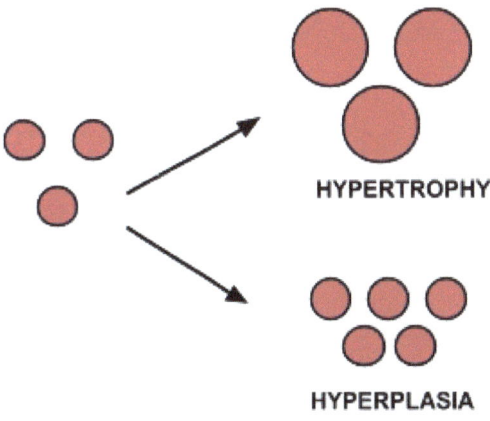

Figure 1: When cells exhibit hypertrophy, they grow larger in size; when cells exhibit hyperplasia, they grow in number (Eriksson-Hogling et. al, 2015).

Hypertrophy in white adipose tissue has been directly correlated to an increased number of proinflammatory adipokines (Andersson *et. al*, 2014). White adipose tissue is heavily involved in energy storage and is therefore associated with obesity, whereas brown adipose tissue produces heat. As mentioned earlier, proinflammatory cytokines, such as tumor-necrosis factor (TNFα), impair triglyceride storage and promote the release of triglycerides as free fatty acids (Eriksson-Hogling *et. al*, 2015). Free fatty acids have been shown to disrupt the insulin signaling pathway; therefore, hypertrophy is strongly linked to insulin resistance in cells (Boyle *et. al*, 2016). By analyzing cells that remain with individuals for the remainder of their lives and observing patterns of hyperplasia or hypertrophy, it is possible to demonstrate how maternal weight during pregnancy can potentially impact patterns of insulin signaling in infants. MSCs are highly beneficial for this purpose as they lead to the development of muscle and white adipose tissue *in utero* and are strong representations of obesity related characteristics that infants may develop later in life (Boyle *et. al*, 2016). Since high maternal BMI during pregnancy has been shown to negatively impact the infant, it is important to understand how maternal BMI specifically affects infant susceptibility to develop metabolic diseases such as type 2 diabetes.

In this study, Ob-MSCs and NW-MSCs were differentiated into white adipocytes, and their growing patterns were analyzed to see if infants born to obese mothers have hypertrophic adipocytes, which indicate susceptibility to becoming insulin resistant. The MSCs were plated in 3D collagen matrices, adipogenesis was induced, and the cells were stained for BODIPY, Wheat Germ Agglutinin, and DAPI; these factors indicated how much lipid the cells stored, how large they were, and how many cells were located in the matrix, respectively. The hypothesis was that adipocytes differentiated from Ob-MSCs would exhibit hypertrophy when compared to those derived from NW-MSCs, indicating that they may also have decreased insulin sensitivity.

METHODS

The MSCs used this research were acquired from Dr. Boyle's lab; Dr. Boyle has utilized umbilical tissue samples to culture MSCs from 164 infants as part of the Healthy Start Baby BUMP Study (R01DK076648; NCT02273297, PI: Dana Dabelea, MD, PhD). The primary goal of Dr. Boyle's study is to analyze how the intrauterine environment impacts infant health. All sample collection was approved by the HIPAA privacy documents of the parent Healthy Start study and Colorado Multiple Institutional Review Board (COMIRB). Since samples incorporated in this study were de-identified, COMIRB classified the research as "non human subject research."

MSCs were collected from the the umbilical cords of 19 infants with mothers identified as obese prior to pregnancy, with BMIs of ≥ 30 kg/m², and 20 infants with mothers classified as normal weight, with BMIs ranging from 19 - 24 kg/m². After the MSCs were cultured, they were differentiated into adipocytes and stained to analyze their growing patterns.

Experimental Design

The growing patterns of the 19 adipocyte samples differentiated from Ob-MSCs were compared to the 20 adipocyte control samples differentiated from NW-MSCs. The cell culture protocols followed, type of media utilized, and the environment the cells were exposed to were kept constant for all cell samples.

All MSCs were cultured in 3D collagen matrices (**Pic. 1**).

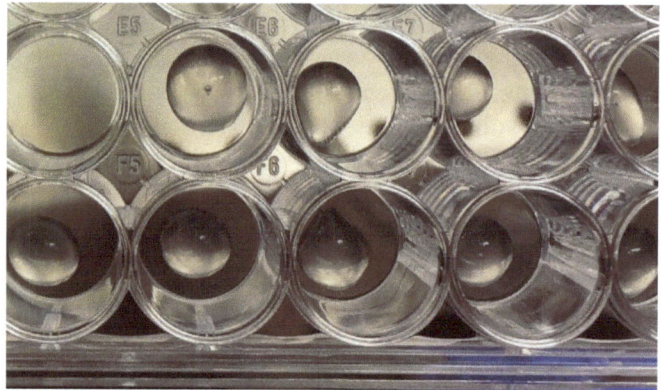

Picture 1: These 3D collagen beads contain cells from a NW-MSC sample; there are approximately 100 MSCs in each bead.

Approximately 100 MSCs from each umbilical cord sample were plated into 24 wells of a 48 well Falcon Corning Cell Bind plate, after which adipogenesis was induced over a 14 day period. The 24 wells for each sample provided more than the used six replicates used for each assay. After the MSCs differentiated into adipocytes, six of the 3D beads were stained with BODIPY, Wheat Germ Agglutinin (WGA), and DAPI and imaged. BODIPY is a neutral lipid stain, and allows for the visualization of lipid content in the cells; WGA is a cell membrane stain, allowing for the quantification of cell size; DAPI is a nuclei stain which allows for cell counting (**Pic. 2**).

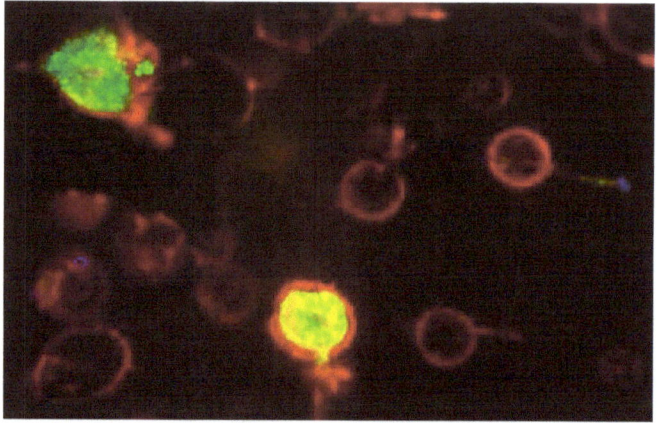

Picture 2: This is a cross section of a 3D collagen bead that contains several adipocytes. The red stain is WGA, which outlines the cell membrane; the green stain is BODIPY, which highlights the neutral lipid content. There is no DAPI stain visible, as the confocal microscope only imaged the very top of the bead. Each image is one slice of the bead, and as the microscope starts imaging the middle of the bead, the fluorescence changes.

Cell Culture and Plating

The MSCs used in this experiment were cryopreserved directly after collection from the infant's umbilical cord. To culture the cells, the MSCs were first thawed in a hot water bath at 37°C and transferred to 100 mm X 20 mm cell treat petri plates with 6 mL mesenchymal growth media, consisting of 500 mL low glucose Dulbecco's Modified Eagle's Medium (DMEM), 25 mL heat inactivated 5% Fetal Bovine Serum (FBS), 5000 ug 1X penicillin /streptomycin (pen/ strep), 50 μL 1 μM Dexamethasone (DEX), 200 μL 0.2 mM Indomethacin (INDO), and 500 μL 0.5 mM 3-isobutyl-1-methylxanthine (IBMX). Cells were then cultured in a CO_2 incubator at 37°C with 5% CO_2. When the MSCs reached 90% confluency, they were trypsinized and plated into 3D collagen matrices (3D Col T-Gel Kit from 101Bio) in 24 wells of the 48 well plate, where adipocyte differentiation was induced. In order to do this, the cells were washed with 4 mL of 1X Dulbecco's Phosphate-Buffered Saline (DPBS), trypsinized with 3 mL of trypsin EDTA at a concentration of 0.05% trypsin and 0.53 mM EDTA, exposed to 3 mL of 10% FBS, and centrifuged at 1000 rpm for 5 minutes. Then, the supernatant was aspirated, 1 mL of mesenchymal growth media was added to the sample, and it was centrifuged once more.

To plate the MSCs into the 3D collagen beads, 600 μL of the cell suspension was transferred to a microcentrifuge tube, centrifuged, and washed one more time with mesenchymal growth media. From there, the excess media was aspirated from the cells and 30.8 μL of component A and 569.2 μL of component B (101 Bio) was added to the cell suspension. Lastly, 40 μL of the cell collagen matrix was placed directly in the center of each well in the plate, ensuring that the bead was not disrupted and did not contain any air bubbles ("3D Col-T Gel," 2016). The 48 well plate was then placed into the CO_2 incubator at 37°C with 5% CO_2. After the cells incubated for 45 minutes, 200 μL of mesenchymal growth media was added to each well, and the cells were further monitored until 70% confluency was reached. Following this, the mesenchymal growth media

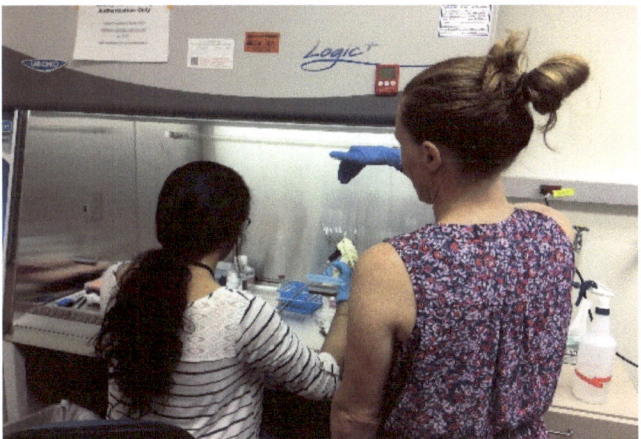

Picture 3: The 3D collagen matrices containing MSCs have been submerged into adipogenesis induction media. This allows for the MSCs to be differentiated into adipocytes; it contains various growth factors such as DEX, INDO, and FBS.

was replaced with adipogenesis induction media, consisting of 500 mL low glucose DMEM, 25 mL heat inactivated 5% FBS, 500 μL 1X pen/strep, 50 μL DEX, 200 μL INDO, 50-0 μL IBMX, and 99 μL of 170 nM insulin, which induced adipo-genesis (**Pic. 3**). Every three days, the media was aspirated and replaced, alternating between adipogenesis induction media and adipogenesis maintenance media, to protect the cells from heavy oxidative stress. The adipogenesis maintenance media consisted of 500 mL low glucose DMEM, 25 mL heat activated 5% FBS, 500 μL 1X pen/strep, and 99 μL 170 nM insulin.

Staining and Imaging

Prior to staining for BODIPY, WGA, and DAPI, the adipocytes were fixed. In order to fix the cells in the 3D culture, the media from the plate was aspirated, ensuring that the collagen bead was not disrupted. Next, with an inoculation loop, the bead was transferred to a microcentrifuge tube and was incubated with the cap off in 500 μL of 10% formalin for 15 minutes at room temperature (**Pic. 4**). Following this, the formalin was removed, and the bead was washed twice with 500 μL of 1X DPBS. Lastly, the fixed cell/collagen matrix was stored in the microcentrifuge tube in 500 μL DPBS at 4°C until staining was conducted.

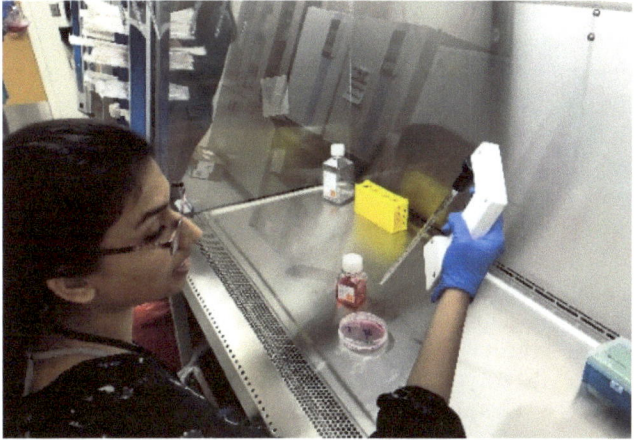

Picture 4: A collagen bead was fixed with 10% formalin. This allowed the adipocytes in the bead to be ready for DAPI, BODIPY, and WGA staining.

In order to stain a fixed 3D collagen bead for BODIPY, WGA, and DAPI, the bead was removed from the 4°C environment and incubated at 37°C for 15 minutes; following this, the bead was permeabilized with 100 μL 0.1% triton and incubated at 37°C for 15 minutes. The triton was then removed, and the collagen bead washed three times with 1X DPBS, with a 5 minute incubation at room temperature between each wash. Next, 3 uM of BODIPY, 3 uM of WGA, and 2 uM of DAPI was added to the 3D bead. Following this, the stained cells were incubated at room temperature for 30 minutes. After the incubation period, the bead was washed twice with 100 uL of 1X DPBS, with a 5 minute incubation at room temperature between each wash. The 3D collagen matrix was transferred to a cover slip and imaged using a 3i Marianas spinning disk confocal microscope with the 20X air objective; the Z Stacks

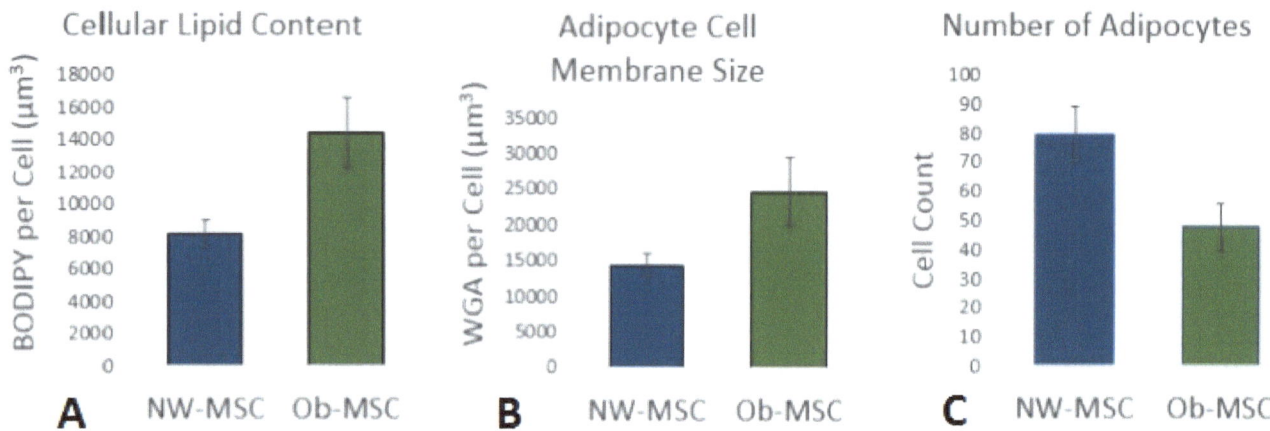

Graph 1: These graphs demonstrate the differences in the lipid content (**A**), cell membrane size (**B**), and cell number (**C**) of adipocytes derived from Ob-MSCs and NW-MSCS. Adipocytes from Ob-MSCs have 77% more cellular lipid content (**A**), are 72% larger in size (**B**), and are 60% fewer in number (**C**).

consisted of 200 images with step size of 1 μm. Following this, the images were analyzed using Fiji, which allowed for quantification of the fluorescence from the cells. For this, a t-test was then utilized to analyze differences in size, quantity, and lipid content between the adipocytes differentiated from the NW-MSC and Ob-MSC samples.

RESULTS
After quantifying the BODIPY stain with Fiji, it was determined that the volume of the neutral lipid content in the adipocytes from Ob-MSCs was approximately two times greater than adipocytes from NW-MSCs. The average volume of the adipocytes from the Ob-MSCs measured as 14373 μm³ with BODIPY which is 77% more lipid content than NW-MSCs, with an average of 8136 μm³ (**Graph 1A**). When evaluated using a t-test, this difference was statistically significant (p = 0.006). Similarly, on average, the WGA stained 24,689 μm³ of an adipocyte from an Ob-MSC, indicating that they were 72% greater in size than adipocytes derived from NW-MSCs, which had 14381 μm³ of WGA (**Graph 1B**). The observed difference in size was statistically significant (p = 0.030). After analyzing the DAPI stains with Fiji, it was determined that there were less adipocytes in the cultures from Ob-MSCs overall in the collagen beads than adipocytes from NW-MSCs. There were 48 cells in the Ob-MSC cultures and 80 cells in the NW-MSC cultures per bead, demonstrating that adipocytes derived from Ob-MSCs were 60% fewer in number (**Graph 1C**). This was statistically significant when analyzed using t-test (p= 0.003).

DISCUSSION
This study analyzed the impact of maternal weight on the growing patterns of adipocytes differentiated from umbilical cord MSCs from infants. Adipocytes either exhibit hypertrophy and grow larger in size or hyperplasia and grow in number; hypertrophy in adipocytes has been associated with release of more proinflammatory cytokines and decreased insulin resistance in adipocytes. Hypertrophy is characterized by greater cell size and lipid content, with fewer number of cells. To assess this, 19 Ob-MSCs

and 20 NW-MSCs differentiated into adipocytes and stained with BODIPY, WGA, and DAPI. In comparison to the control group, the adipocytes differentiated from Ob-MSCs had 77% more lipid content, were 72% larger, and were 60% fewer in number than those from NW-MSCs. Prior to this study, it was unclear whether higher fat mass percentage in infants born to obese mothers was due to hypertrophy or hyperplasia. The staining and imaging results of this investigation confirmed the hypothesis that adipocytes differentiated from Ob-MSCs would exhibit hypertrophy, not hyperplasia. The t-test that was conducted on these results demonstrated statistical significance between the adipocytes derived from Ob-MSCs and those from NW-MSCs for BODIPY (p = 0.006), WGA (p = 0.030), and DAPI (p = 0.003).

Hypertrophy in adipocytes differentiated from Ob-MSCs may, in part, explain why infants of obese mothers have a heightened risk for developing insulin resistance later in their life (Stephens, 2012). Individuals with white adipose tissue hypertrophy have demonstrated significantly higher levels of proinflammatory cytokines; these cytokines initiate the metabolism of triglycerides into free fatty acids, which disrupt the receptor proteins that initiate the insulin signaling pathway (Andersson et. al, 2014). Therefore, white adipose tissue hypertrophy has been directly correlated with high insulin resistance (Eriksson-Hogling et. al, 2015). If adipocytes from Ob-MSCs are exhibiting signs of hypertrophy in the fetus itself, infants may start exhibiting decreased insulin sensitivity.

This research, however, needs to be verified with other studies to directly establish a link between maternal obesity and insulin resistance in infants. While the 3D collagen matrices in this study allowed the adipocytes to be more morphologically and physiologically representative of an *in vivo* environment, it also cannot be fully confirmed whether the adipocytes from the Ob-MSCs were fewer because they were larger and a smaller number of them fit into the collagen bead or if there were actually fewer.

The next step for establishing a link between maternal obesity and infant insulin resistance is to conduct an insulin dose response assay on the adipocytes from both groups and analyze proteins associated with the insulin signaling pathway using a Simple Western machine. In an insulin dose response assay, the adipocytes will be exposed to insulin for a set period of time, and the major proteins in the insulin transduction pathway, pAkt and IRS-1, will be quantified (**Fig. 2**). If the adipocytes from Ob-MSCs have a lower amount of these key proteins that allow for cellular uptake of insulin in comparison to those from NW-MSCs, insulin resistance will be further confirmed (White, 2003). This is because the adipocytes will be less receptive to the insulin and, therefore, will produce less of these two key signaling proteins in response to it.

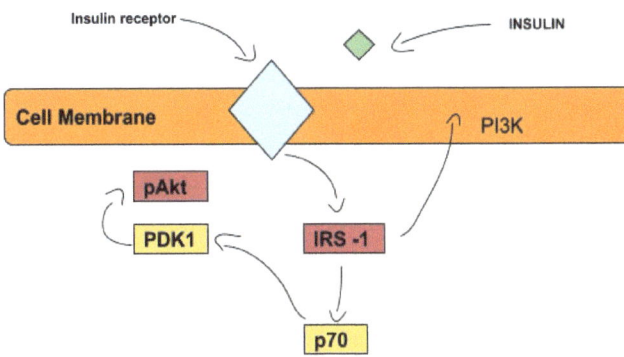

Figure 2: The following image depicts a simplified version of the insulin transduction pathway. pAkt plays a major role in glucose metabolism, and IRS-1 is a receptor protein that marks the beginning of the signaling pathway. These proteins are easily measured with a Simple Western machine (White, 2003).

If infants born to obese mothers have decreased insulin sensitivity, they are at much higher risk of developing type 2 diabetes (Stephens, 2012). This research will better inform women on how choices made before and during pregnancy can affect their future children. The results found can allow doctors to design diet and lifestyle routines that will help women ensure a healthy future for their baby.

ACKNOWLEDGMENTS

I would like to thank Shawndra Fordham and Susanne Petri, my current research and biotechnology instructors, for offering me advice and supporting me in the design of my research project. Specifically, I would like to thank Shawndra Fordham for editing my proposal and paper, helping design and format my scientific poster, and helping me plan my experimental design. Furthermore, I appreciate Bryan Winkelman for helping ensure that my research was effectively communicated to a wider audience by designing my biotechnology website, helping publish my article, and helping design my poster. I am also grateful for Radu Moldovan and Carol Mirita for training me on how to properly use the 3i Inverted Marianas Spinning Disk confocal microscope. Lastly, I would like to thank Anschutz Medical Campus for providing me with lab space, and Dana Dabelea for funding all materials and equipment.

REFERENCES

3D Col-T Gel. (2016). Retrieved from http://www.101bio.com/P720_3D_cell_culture_gel.php

Andersson, D. P., Hogling, D. E., Thorell, A., Toft, E., Qvisth, V., Näslund, E., ... & Dahlman, I. (2014). Changes in subcutaneous fat cell volume and insulin sensitivity after weight loss. *Diabetes care*, 37 (7), 1831-1836. https://doi.org/10.2337/dc13-2395

Boyle, K. E., Patinkin, Z. W., Shapiro, A. L., Bader, C., Vanderlinden, L., Kechris, K., .. .Davidson, E. J. (2017). Maternal obesity alters fatty acid oxidation, AMPK activity, and associated DNA methylation in mesenchymal stem cells from human infants. *Molecular Metabolism*, 6(11), 1503-1516. https://doi.org/10.10-16/j.molmet .2017.08.012

Boyle, K. E., Patinkin, Z. W., Shapiro, A. L., Baker, P. R., Dabelea, D., & Friedman, J. E. (2016). Mesenchymal stem cells from infants born to obese mothers exhibit greater potential for adipogenesis: the healthy start BabyBUMP project. *Diabetes*, 65(3), 647-659. doi: 10.2337/db15-0849

Centers for Disease Control and Prevention. (2018). Overweight & Obesity. Retrieved from https://www.cdc.gov/obesity/data/adult.html

Eriksson-Hogling, D., Andersson, D. P., Bäckdahl, J., Hoffstedt, J., Rössner, S., Thorell, A., ... & Rydén, M. (2015). Adipose tissue morphology predicts improved insulin sensitivity following moderate or pronounced weight loss. *International Journal of Obesity*, 39(6), 893-904. https://doi.org/10.1038/ijo.2015.18

Rosenberg, E. D., & Spiegelman, B.M. (2006). Adipocytes as regulators of energy balance and glucose homeostasis. *Nature, 444*(7121), 84 7-859. doi: 10.1038/nature05483

Spiegelman, B. M., & Flier, J. S. (1996). Adipogenesis and obesity: rounding out the big picture. *Cell*, 87(3), 377-389. https:/ /doi.org /1 0.1016/S0092-8674(00)81359-8

Stephens, J. M. (2012). The fat controller: adipocyte development. *PLoS biology*, 10(11), e1001436. https://doi.org /10.1371 /journal.pbio.1 001436

White, M. F. (2003). Insulin signaling in health and disease. *Science, 30* 2(5651), 1710-1711. doi: 10.1126/science.1092952

ABOUT THE AUTHOR

Pictured: This picture is of Shubhangi along with Dr. Keleher in lab. Each day that Shubhangi left RCHS to go work on campus she took a picture of her and her mentor in lab for attendance. They made it fun each time.

Shubhangi is a senior at Rock Canyon High School who is extremely excited to pursue a degree in biomedical engineering. Shubhangi became a member of the Rock Canyon Biotechnology program to develop skills, such as handling lab equipment and learning how to interpret data, that will be applicable to future bioengineering research careers. Her research idea was inspired by her lab experience in Dr. Kristen Boyle's lab, where she read an article on the strong correlation between hypertrophy and decreased insulin resistance which led to type II diabetes. When she's not "nerding out" and doing homework or working in the Boyle lab, she likes to hike, swim, and read.

While designing her experiment, Shubhangi learned how to troubleshoot problems quickly; during the imaging and staining process, she immediately learned that the stock the stains could not penetrate the bead. From there, she realized that she needed a depth friendly microscope. There were several instances where Shubhangi encountered problems similar to this, and through this process, she learned how to be resilient and rethink how to approach her methods. Overall, this was an amazing experience for her, as she was able to take charge of her own project.

Engineering CRISPR-Cas9 to disrupt the function of the *tnaA* gene and subsequent tryptophanase production in HME63 *Escherichia coli*

M. W. Deckerman, A. M. Deschane, J. C. Stohs & S. L. Fordham
Department of Science, Principles of Experimental Design in Biotechnology, Rock Canyon High School, Highlands Ranch, Colorado, USA

Indole is a compound produced by bacteria that is responsible for the fecal smell associated with *Escherichia coli*. Indole production has also been linked to biofilm formation and the development of colony antibiotic resistance. In *E. coli*, indole is created as a result of the breakdown of tryptophan by tryptophanase, which is produced by the tryptophanase gene (*tnaA*). This project aimed to use the gene editing technology, CRISPR, to disrupt the *tnaA* in *E. coli*, thereby stopping indole production and creating a biological tool to help study biofilm formation and antibiotic resistance. To do this, we engineered a Cas9 plasmid containing a chloramphenicol resistance gene sequence and an arabinose promoter along with a guide RNA (gRNA) sequence corresponding to the *tnaA* gene. We also engineered a single-stranded DNA oligo as the homology directed repair template (HDRT) that would be used by the CRISPR-Cas9 system to repair the double stranded breaks (DSB) following Cas9 editing. The next steps of this project are to clone the engineered pCas9 plasmid and transform it into HME63 *E. coli* along with the HDRT. The CRISPR system can then be initiated to edit the *tnaA* gene by adding arabinose. To test for successful CRISPR editing, an indole test will be performed on individual colonies. If the *tnaA* gene is successfully disrupted, tryptophanase and subsequent indole will not be produced.

Clustered Regularly Interspaced Short Palindromic Repeats (CRISPR) was originally identified in bacteria in the 1980s; however, its function and future significance were unknown at that time (Ishino, Shinagawa, Makino, Amemura, & Nakata, 1987). In 2005, researcher Francisco Mojica with the University of Alicante - Spain correctly hypothesized that the CRISPR system served as an adaptive immune system in bacteria and was used by the cell to target and cleave invading viral RNA or DNA (Ran *et al.*, 2013). Following years of research by many different scientists worldwide, Jennifer Doudna and her team at the University of California - Berkeley demonstrated how CRISPR and the Cas9 protein could be used along with a gRNA to target and cut specific DNA sequences in the prokaryotic genome. Later in 2013, Fen Zhang and his team at the Broad Institute were the first to successfully use the CRISPR/Cas9 system to edit the genome in eukaryotes (Cohen, 2017).

CRISPR works by using a nuclease to create double-stranded breaks (DSB) in the DNA and then exploits the homology directed repair (HDR) pathway, which is already present in the cell, to allow researchers to control how those breaks are repaired. There are three vital components in the CRISPR/Cas9 system: the Cas9 nuclease, a gRNA that is engineered to target a specific cut site, and the HDRT (**Fig. 1**). In prokaryotes, to use the CRISPR/Cas9 system to perform gene editing, a plasmid is often inserted into the cell containing the Cas9-gRNA complex along with the HDRT which can be inserted as a DNA oligo or in a plasmid. The CRISPR RNA (crRNA) and transactivating CRISPR RNA (tracrRNA) form the gRNA complex. This gRNA has been specifically programmed to bring the Cas9 nuclease to the

desired cut-site in the genome where it then creates the DSBs. The tracrRNA is required for crRNA maturation, while the crRNA guides the Cas nuclease to the target sequence. Once at the cut site, the Cas9 nuclease must recognize a specific protospacer adjacent motif (PAM), an approximately 2-6 base pair sequence following the target sequence, or it will not bind to and cut the DNA. Once the DNA has been cut, the HDRT is used by the cell to repair the DSB using its innate HDR system. In HDR, the cell's repair mechanism attempts to repair any DSBs in the DNA using homologous sections of DNA as a guide. When used in

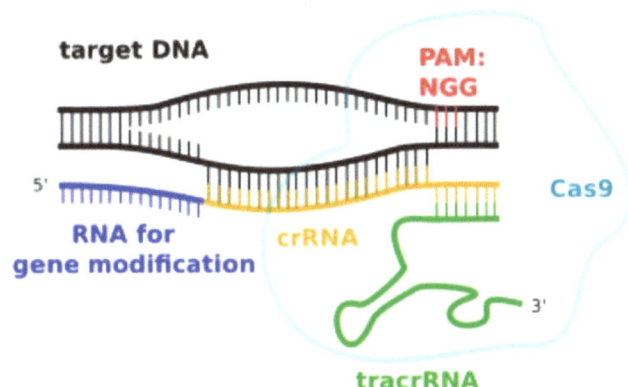

Figure 1: The above figure illustrates the components of the CRISPR/Cas9 system which includes the Cas9 enzyme and the gRNA, a combination of the tracrRNA and crRNA, which forms a complex matching the desired sequence in the DNA. Adapted from "CRISPR-Cas9 mode of action," by V. Anshelm, 2015 (https://commons.wikimedia.org/wiki/File:CRISPR-Cas9_mode_of_action.png). By Wikimedia Commons, 2015. Public domain.

genome editing, numerous repair templates are introduced into the cell that contains the gene or insertion of interest flanked by sequences of DNA that are homologous to the sequence found at each end of the DSB to increase the chance that the cell will incorporate the repair template after the DSB has been created (Addgene, 2018).

CRISPR gene editing technology is being used in disease research and treatment, synthetic biology, agriculture, engineering new antimicrobials, conducting genome-wide screens, and even controlling insect spread of disease. This new gene editing technology has changed biomedical research drastically and is being used to successfully edit everything from bacteria to vertebrates and, more recently, human cells and embryos. CRISPR has previously been used in bacteria to produce virus immune bacteria. Using this knowledge of the capabilities of CRISPR, it was later used in invertebrates, more specifically mosquitoes. *Anopheles gambiae* mosquitoes, the chief transmitters of malaria, were genetically altered to have the FREP1 gene disrupted as it helps malaria to survive in the gut of the mosquitoes and develop to the stage where it can be transmitted into humans. This research could potentially help reduce the transmission of malaria from mosquitoes to humans (Barrangou & Horvath, 2010). More recently, CRISPR research has started in humans. The first human clinical trial using CRISPR technology was approved in China, on July 6th, 2016. It had the goal of treating metastatic non-small cell lung cancer using the CRISPR-Cas9 complex to target the gene coding for the PD-1 protein, which checks on the cell's capacity to launch an immune response. The patient's cells were edited *in vitro* and reintroduced to the body with the goal of allowing the T-cells to more effectively stop the rapid growth of the cancer cells (Cyranoski, 2016).

Our engineering project sought to use the CRISPR/Cas9 system to silence the *tnaA* gene in the HME63 strain of *E. coli* and halt the production of tryptophanase. In *E. coli,* the *tnaA* gene is responsible for the production of tryptophanase, which breaks down tryptophan and water to produce indole, ammonia, and pyruvic acid **(Fig. 2)**. Indole is the compound that creates the undesirable fecal smell commonly associated with many strains of bacteria. Engineering the CRISPR system to stop the production of tryptophanase and subsequent production of indole would create a tool that could be used in industry and research that has a lowered offensive odor. In 2006, as part of the International Genetically Engineered Machine competition,

MIT students engineered *E. coli* to produce wintergreen and banana scents instead of the foul indole scent using only endogenous metabolites already present in the bacteria. Instead of using CRISPR and genome editing, this team transformed bacteria with plasmids that would produce enzymes that converted chemicals native to the bacteria into the aromatic compounds of methyl salicylate and isoamyl acetate, the scents of wintergreen and banana, respectively (Venkatachalam *et al.*, 2006). While the goal of this research, to lower the indole production, was similar to ours, we strove to create a CRISPR genome edited organism completely lacking in tryptophanase and subsequent indole that could then be further modified to best meet the needs of the individual research team.

Indole is a biologically relevant compound that has been linked to biofilm formation, plasmid stability, and antibiotic resistance in *E. coli*. It has been found to be an extracellular signaling molecule that *E. coli* uses to regulate growth and division and perform quorum sensing and it positively impacts biofilm formation in pathogenic strains of *E. coli* (Jamal *et al.*, 2018). Bacterial biofilms are linked with disease and are known to be highly antibiotic resistant (**Pic. 1**). Over half of all microbial infections and 80% of all hronic infections have been linked to biofilms (Jamal *et al.*, 2018). Engineering an indole negative model would be useful to researchers in investigating the mechanisms involved with antibiotic resistance and biofilm production as a result of indole. Biofilms in bacteria pose a large threat to human

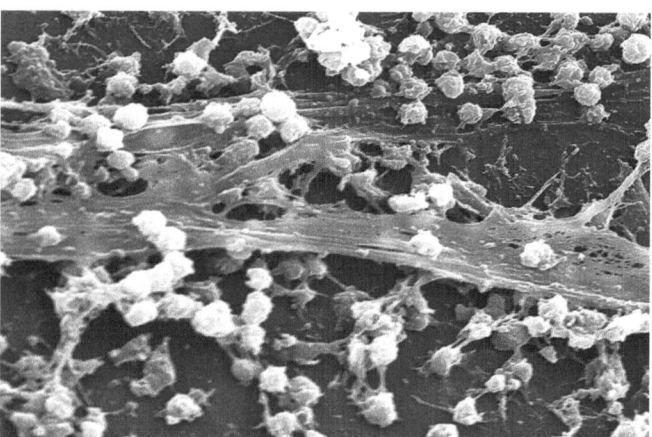

Picture 1: The above image shows *Staphylococcus aureus* biofilms. From *staphylococcus aureus* biofilm, CDC/ Rodney M. Donlan, Ph.D.; Janice Carr (PHIL #7488), 2005. Public domain.

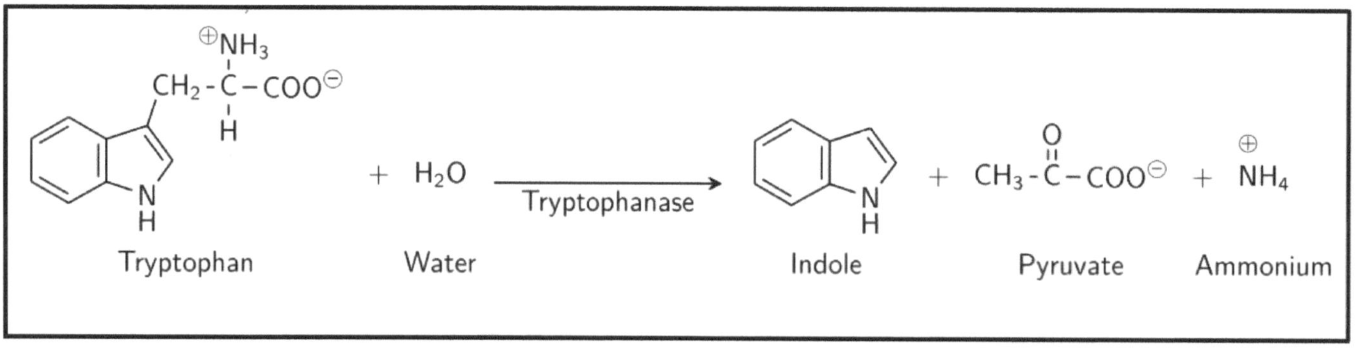

![Figure 2 chemical reaction diagram]

Tryptophan + H₂O → (Tryptophanase) → Indole + Pyruvate + Ammonium

Figure 2: The above diagram illustrates the chemical breakdown of tryptophan into indole by tryptophanase. From "Indole Synthesis" by Krishnavedala, 2014 (https://en.wikipedia.org/wiki/File:Indole.svg). Public domain.

health as they are involved in a variety of infectious diseases and are less recognized by the immune system. In order to develop methods to prevent the development of biofilms, a greater expanse of knowledge on them is needed (Lee & Lee, 2010). Stopping the production of indole in a bacterial population may make *E. coli,* and other indole producing bacteria, more susceptible to antibiotics and easier to treat in disease models. Additionally, working with CRISPR edited bacteria in the laboratory could make antibiotic screens more effective.

Prior research has indicated that tryptophanase is synthesized by *E. coli* solely to break down tryptophan into indole, so a lack of tryptophanase should not alter cell growth (Li & Young, 2013). The subsequent build-up of tryptophan is controlled by the trp operon in a negative feedback loop in which excess tryptophan binds to the trp repressor and prevents transcription of the trp gene, thereby stopping further tryptophan production (Bhagavan & Ha, 2002). Once this project is completed and the CRISPR system is induced to stop all indole production, these mechanisms will be tested. *E. coli* has become a crucial tool in modern biotechnology and is widely used in many different applications. It is also easy and fast to grow, as it thrives in anaerobic and aerobic conditions. *E. coli* has a well understood genome and is industrially useful. Its versatility has resulted in it being studied intensively all across the world; it is now one of the best understood organisms scientists work with today. We worked with the HME63 strain of *E. coli* due to its inducible bacteriophage lambda red recombination function that is under the control of a temperature sensitive repressor. Lambda red is the cell's natural DNA repair mechanism; without it the cell would die when CRISPR cleaves the DNA. In addition to the lambda red pathway, HME63 also has a disabled mismatch repair system which means that it cannot easily identify if modifications or mutations to the genome have happened, which allows genome editing to be more efficient. HME63 also has a built-in ampicillin (Amp) resistance which helps reduce possible contamination in culture and provides an effective screening tool for selection of HME63 colonies.

We engineered the pCas9 plasmid with the gRNA sequence matching the *tnaA* gene as well as the HDRT that will be used to repair the DSB. Once transformed into HME63 *E. coli* along with the HDRT, the CRISPR/Cas9 system should target the *tnaA* gene ultimately stopping the production of tryptophanase and subsequent indole. Future steps of this research include cloning the engineered plasmid, using electroporation to introduce the engineered plasmid and HDRT into HME63, and performing an indole test on cultures of individual colonies. If indole is not produced in the engineered bacteria, but is present in the non-CRISPR edited control, it could be concluded that the gene editing was successful. The experiment should then be repeated three times with multiple trials and transformation and the resulting CRISPR efficiencies measured. To further verify successful genome editing, DNA sequencing should be performed on the *tnaA* gene for both the edited and control bacteria. This project is the first step necessary to engineer a useful tool for scientists to use in scientific research while also further exploring the applications of CRISPR technology.

METHODS/RESULTS
We worked towards engineering the CRISPR-Cas9 system to disrupt the *tnaA* gene and stop the subsequent production of tryptophanase in the HME63 strain of *E. coli*. Successful completion of this project required several key steps:
- cloning and miniprep of the pCas9 plasmid;
- performing restriction digest, purification, and ligation of the pCas9 plasmid with the previously engineered gRNA oligo;
- engineering the homology directed repair template (HDRT);
- cloning and miniprep of the engineered pCas9 plasmid;
- transforming HME63 *E. coli* with the engineered pCas9 plasmid and engineered HDRT;
- inducing CRISPR-Cas9 in HME63 *E. coli*;
- and testing cultures of individual colonies of HME63 for the presence of indole.

pCas9 Cloning and Miniprep
In the first stage of our project, we cloned the pCas9 plasmid **(Fig. 3),** obtained from AddGene, using the DH5α strain of *E. coli*. The pCas9 plasmid is a low copy plasmid, meaning we should expect to yield 0.2-1 μg/mL of DNA compared to a high copy plasmid in which we would expect to yield 3-5 μg/mL of DNA (Growth of Bacterial Cultures, n.d.). We then performed a miniprep to purify the pCas9 plasmid following the protocols outlined in the Thermo Fisher Miniprep Kit K0502. The bacteria were cultured in LB with 25 μg/mL chloramphenicol (Cam) at 37°C **(Pic. 2)**. To miniprep the pCas9 plasmid out of the *E. coli,* the bacteria were pelleted, resuspended, and lysed with 250 μL of resuspension solution and 250 μL of lysis solution. 350 μL of neutralization

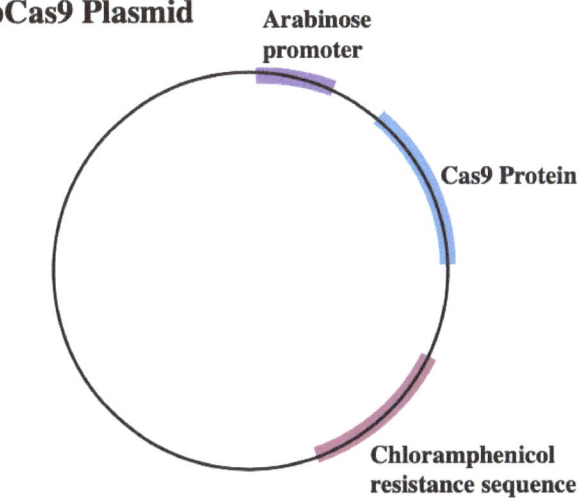

Figure 3: The image above depicts the engineered pCas9 plasmid which includes a chloramphenicol resistance sequence, the sequence for the Cas9 Protein, and an arabinose promoter. The Cas9 protein sequence will only be expressed in the presence of arabinose due to the promoter.

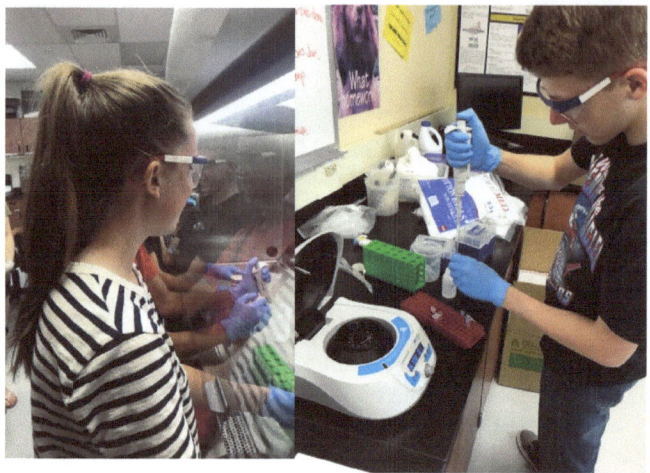

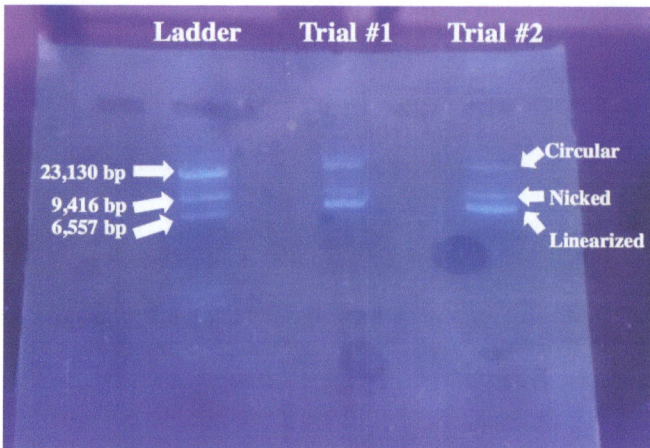

Picture 4: The above image shows our successful gel electrophoresis after restriction digest. Two trials were run through the gel alongside a λ DNA-HindIII Digest Ladder from NEB. The gel separates circular (no cuts), nicked (one cut), and linearized (two cuts) plasmid after restriction digest. The linearized plasmids were cut from the gel and purified for use later.

Picture 2: Here we are making liquid cultures in the bio-logical safety cabinet to pre-pare for miniprep of the pCas9 plasmid.

Picture 3: In the picture above, we are performing a miniprep to isolate the pCas9 plasmid from DH5α bacteria to prepare for restriction digest.

solution was added and the sample was centrifuged at 16,100 rpm for 5 minutes. The nucleic acids were then separated from the solution using a spin column which was further washed with DNA wash solution twice and the flow through discarded. Next, the pCas9 plasmid was eluted from the column with 35 μL of elution buffer, composed of 10 mM Tris-Cl at a pH of 8.5 (**Pic. 3**). The sample was centrifuged and the flow through collected this time contained the purified pCas9 plasmid. To quantify the yield of the minipreps, we used a NanoDrop 2000C from Thermo Fisher Scientific. Our minipreps yielded low yet sufficient concentrations of 96.1 ng/μL and 104.8 ng/μL pCas9 plasmid.

Restriction Digest, Purification, and Ligation
In order to insert the gRNA sequence into the pCas9 plasmid, we performed restriction digest and ligation on the purified plasmid following the New England Biolabs protocols (NEBcloner, n.d.). To do this, we added 1.5 μL of 10X CutSmart Buffer and 1 μL BsaI enzyme, donated by New England Biolabs (NEB), to 12.5 μL of the purified pCas9 plasmid solution and incubated the solution in a hot water bath at 37°C for 12 hours. We then performed gel electrophoresis to separate the circular plasmids (no cuts), nicked plasmids (one cut), and the linearized plasmids (two cuts) (**Pic. 4**). We also used a λ DNA-HindIII Digest Ladder from NEB, which has bands ranging from 125 bp to 23,130 bp. Since our smallest band should have been approximately 9000 bp and the ladder has a band that is 9,416 bp long, we were able to identify the linearized plasmids. To purify this plasmid from the gel, we cut the 9000 bp band, which contains the linearized plasmids, from the gel, and using a Gel DNA Recovery kit from Zymo Research, we purified the plasmid from the agar. To do this, the agarose was dissolved in dissolving buffer at a 3:1 agarose gel to dissolving buffer ratio. The solution was incubated at 55°C until the gel was completely dissolved, after which it was transferred to a Zymo-Spin Column and centrifuged at 6000 rpm for 60 seconds. The spin filter was washed with 200 μL of DNA

wash buffer then centrifuged for 30 seconds. The flow through was discarded and the DNA was eluted from the spin column using 6 μL DNA elution buffer and added directly to the column matrix. This left the purified, digested pCas9 plasmid.

After successfully purifying the digested pCas9 plasmid, we ligated it with the gRNA oligo, which was previously designed by last year's CRISPR research students, Peter Deckerman and Jonathan Ferry, using Benchling. The sequence for this oligo is as follows: 3'TATTTCATCAAGCAGCGTGAAGCAGAATACAAA GACTGGACCATCGAGCAGATCACCCGCGAAACCT ACAAATAT5' We have confirmed, using Benchling, that the gRNA sequence is specific to the *tnaA* gene in HME63 and as such should limit any off target effects. To ligate the plasmid with the gRNA oligo, digested pCas9 plasmid was mixed with the gRNA oligo in a 1:3 ratio. The ligation reaction consisted of 0.5 μL of the plasmid: gRNA mixture along with 1 μL 10X ligase reaction buffer, 1 μL ligase, and

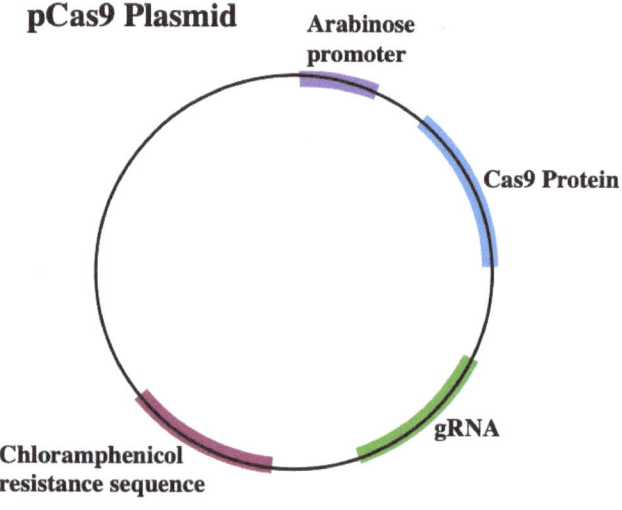

Figure 4: The image above depicts the engineered pCas9 plasmid with the ligated gRNA that corresponds to the *tnaA* gene.

11.5 μL of heat treated distilled water made to destroy any extraneous DNA. The mixture was incubated at room temperature for 2 hours and heated at 65°C for 10 minutes. After that was completed, we stored our engineered pCas9 plasmid at 4°C. The solution contained an average of 50 μg/mL plasmid containing the genes for the pCas9 protein, chloramphenicol resistance, and the gRNA sequence (**Fig. 4**).

Homology Directed Repair Template

One important step of this project, before performing transformation of the HME63 *E. coli* with the engineered plasmid, was to design a custom HDRT. To do this, we used the website Benchling which allowed us to select the *tnaA* sequence and enter our desired mutation, the stop codon ACT. Benchling then allowed us to mutate our gRNA sequence so the Cas9 protein does not cut our HDRT. Lastly, we entered in the desired length of our homology arms to complete the design of the HDRT. The following sequence is the HDRT sequence that we engineered:
5'AGGTGGTCAGCCGGTTTCACTGGCAAACTTAAAA GCGATGTACAGCATCGCGAAGAAATACGATATTCC GGTGGTAATGGACTCCGCGCGCTTTGCTGA**ACT**AA ACGCCTATTTCATCAAGCAGCGTGAAGCAGAATAC AAAGACTGAACCATCGAGCAGATCACCCGCGAAA CCTACAAATATGCCGATATGCTGGCGA3'.
After engineering this HDRT, we purchased it from Integrated DNA Technologies.

Miniprep of Engineered pCas9 Plasmid

Before we could induce CRISPR, we needed a high concentration of the engineered pCas9 plasmid containing the gRNA sequence. To do this, we first attempted to electroporate our engineered pCas9 into DH5α *E. coli*. We prepared for electroporation by making LB broth (no antibiotics), culuturingDH5α on LB/Cam plates, and sterile double distilled water (ddH$_2$O). To make competent bacteria, we washed a 2 mm colony of DH5α three times by resuspending the bacterial mass in 1 mL of ddH$_2$O and centrifuging the solution at 6000 rpm for 5 minutes at 4°C. Once washed, we removed the supernatant and resuspended in 40 μL of ddH$_2$O. To transform our plasmid into the DH5α *E. coli*, we added 1 μg of plasmid into our 40 μL bacterial suspension and transferred the suspension into a 2 mm gap BTX Electroporation Cuvette Plus. The cells were then electroporated in a BTX Gemini X² Electroporator utilizing an exponential decay with the following settings: 1.8 kV, 5 ms, 200 ohms, and 25 μF (**Pic. 5**). After electroporation, we recovered the cells in 1 mL of LB broth without antibiotics at 37°C in the shaking incubator at 200 rpm. After 20 minutes, 100 μL of the liquid culture was spread onto LB agar plates containing chloramphenicol at concentrations of 25 μg/mL. The plates were then incubated at 37°C overnight.

This process was repeated twice with both times failing to result in colony growth, signifying that the transformation of our engineered pCas9 plasmid did not successfully occur. One possible explanation we found after discussing our results with other research scientists was that our wash and

suspension of the bacteria was in water which can cause osmotic problems for the cells. In the future, it was recommended that we use a 10% glycerol solution in place of ddH$_2$O as this will be an isotonic environment for the cells.

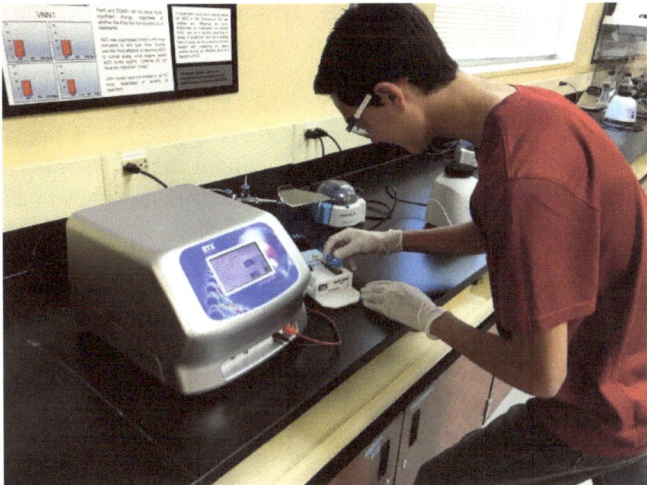

Picture 5: In the above picture we are performing electroporation on DH5α in order to transform the engineered pCas9 plasmid.

FUTURE STEPS

To successfully complete this project, the engineered plasmid needs to be successfully cloned and along with the HDRT transformed into HME63 *E. coli* and the CRISPR system induced. Then, subsequent indole production of individual colonies needs to be tested. Due to time constraints, we were unable to complete these steps to finish this project. The critical next steps are outlined below.

Inducing CRISPR

After repeating the miniprep of the engineered plasmid in DH5α *E. coli*, the engineered plasmid along with the HRDT needs to be electroporated, following the protocols outlined earlier with the suggested revisions, into the HME63 strain of *E. coli*. We recommend using the electroporation protocols for the BTX Gemini X² Electroporator machine utilizing an exponential decay with the following settings: 1.8 kV, 5 ms, 200 ohms, and 25 μF. Prior to electroporation, the lambda red pathway needs to be activated by culturing the HME63 cells at 42°C for 15 minutes before making the cells electrocompetent. After electroporation, the cells should be plated on LB/Cam/Arabinose (Ara) plates that were prepared with 25 μg/mL Cam and 200 mg/mL Ara. Successfully transformed cells should have received the plasmid which infers resistance to chloramphenicol, while arabinose is required to turn on the CRISPR system in the cells. In addition to the treatment plate containing HME63 cells that had been transformed with the engineered plasmid and the HDRT, the following control plates should also be made. First, to test the efficacy of the chloramphenicol, non-transformed HME63 cells should be plated on a LB/Cam/Amp plate. We expect no growth on this plate. If a lawn of growth appears, we will know that Cam is not working. Next, to verify that the bacteria are alive after electroporation, non-transformed but electroporated HME63 should be plated on a LB/Amp plate. This plate should have a thin lawn of growth. If there is no growth on this plate it will indicate a problem with the bacteria (**Table 1**).

CRISPR Verification

To verify successful CRISPR genome editing of the *tnaA* gene, individual colonies should be selected from the +Plasmid +HDRT LB/Cam/Amp/Ara plate and an overnight culture made of each individual colony in LB broth containing 25 µg/mL chloramphenicol. After 18-28 hrs of incubation at 37°C, each individual culture should be initially screened using an olfactory test to determine if the cells lack the characteristic fecal smell associated with *E. coli*. Following this olfactory screen, an indole test should be performed on cultures that indicate possible success with CRISPR genome editing. The indole test will ensure that indole production has been stopped by testing the reaction between indole and dimethylaminobenzaldehyde which results in a visible reddish-pink color change in the presence of indole (MacWilliams, 2009). To perform the indole test, a colony of *E. coli* is inoculated in a tryptone broth and incubated at 35°C for 24-48 hours. After culturing, 4-5 drops of Kovac's reagent, which contains dimethylaminobenzal-dehyde, is added to the culture tube. A positive result, as indicated by a reddish-pink color, confirms the presence of indole indicating that CRISPR editing was unsuccessful in silencing the *tnaA* gene for this colony. No color change is indicative of the lack of indole production and suc-cessful CRISPR genome edit-ing for this particular colony (**Pic. 6**). An indole test per-formed on a colony that was not transformed should serve as a control to ensure the test works properly. Following the indole test, successful *tnaA* gene disruption also should be confirmed by amplifying the *tnaA* gene using PCR and performing Sanger sequenc-ing on the amplicons.

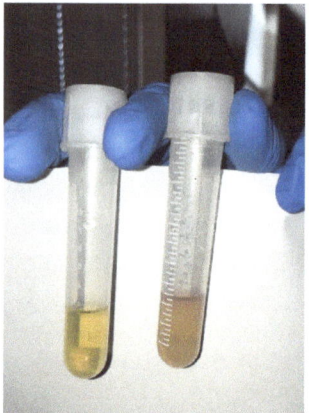

Picture 6: The picture above depicts an Indole test. On the left is a negative result and on the right is a positive result.

DISCUSSION

CRISPR is becoming increasingly more relevant in the biotechnology field, along with the need for antibiotic resistance and biofilm formation study. As mentioned earlier, biofilms are highly antibiotic resistant and are found in approximately 50% of all microbial infections and 80% of all chronic infections (Lee & Lee, 2010). The goal of this project was to use CRISPR/Cas9 to disrupt the *tnaA* gene in the HME63 strain of *E. coli* and subsequently stop indole production as tryptophanase breaks down tryptophan and produces indole, pyruvate, and ammonium. This engineered HME63 strain of *E. coli* would serve as a tool to study biofilm formation along with antibiotic resistance as both are increasingly growing problems the world faces.

We successfully completed many of the steps required to create this engineered strain of *E. coli*. We cloned and mini prepped the pCas9 plasmid using DH5α *E. coli* with high plasmid concentrations of 104.8 ng/µL and 96.1 ng/µL which is near the minimum concentration of 100 ng/µL for a successful miniprep of a low copy plasmid. We also digested this plasmid using the BsaI restriction enzyme, performed gel electrophoresis, and purified it from the agarose. We then successfully ligated the gRNA sequence corresponding to the *tnaA* gene into the pCas9 plasmid. Another critical step we accomplished in this research was designing the HDRT engineered with a stop codon between the homology arms. We ordered this HDRT from IDT and stored it at -20 °C for future use. Unfortunately, time constraints inhibited us from completing the last steps of this research project, consisting of transforming the engineered plasmid into DH5α, cloning and mini prepping the engineered plasmid using DH5α *E. coli*, transforming both the engineered plasmid and the HDRT into HME63 *E. coli*, inducing the CRISPR system, and confirming its success by performing an indole test.

We cannot determine where we were unsuccessful due to our inability to verify the success of each step. We may have lost large amounts of plasmid while performing a restriction digest due to the instability of the linear DNA which could have degraded in a quicker manner. Due to low plasmid concentrations, we may have had too few plasmids to successfully complete the steps that occur after. If restriction digest was successful, we may have unsuccessfully ligated our gRNA into pCas9 as head-to-head and head-to-tail may have occurred without our knowledge. If we successfully engineered the pCas9, then electroporation could be the last

Controls Needed to Screen for Successful Plasmid Transformation		
Plate 1	+Plasmid +HDRT on LB/Cam/Ara	Experimental plate used to screen for plasmid transformation
Plate 2	-Plasmid -HDRT on LB/Cam/Amp	Control plate used to test efficacy of chloramphenicol
Plate 2	-Plasmid -HDRT on LB/Amp	Control plate used to test efficacy of electroporation

Table 1: The above table shows the three different plates that will be used to observe the bacteria. Plate 1 will be used to screen for successful plasmid transformation as only the bacteria that receive the plasmid will be able to grow in the presence of Cam, and the Ara will allow for the Cas9 protein to be expressed, so we would expect gene editing on any colonies that form. Plate 2 will allow us to see the efficacy of the Cam we will be using, as no bacteria should be able to grow due to the lack of plasmid transformation. Plate 3 will test to see if the *E. coli.* we are using will survive the electroporation protocols.

unsuccessful step. As described earlier, ddH₂O created a harsh environment for the bacteria and could have resulted in a decreased amount of cells that survived the electroporation process.

Other researchers have struggled with restricting and ligation of plasmids, a critical step to the CRISPR pathway (**Pic. 7**). Researchers who electroporated a linearized plasmid to lead to the formation of transgenic plants described that electroporation was extremely efficient compared to free-DNA transfer. However, they found that plasmid ligation appeared to be random, with head-to-head and head-to-tail ligated plasmids. In addition to this, difficulties also occurred while restricting their plasmid. Some of their electroporated plants displayed rearrangements or modifications of the transformed plasmid thought to have occurred due to the linearized shape of the plasmid. The plasmid may have undergone rearrangement after it was transformed into the cell (Riggs & Bates, 1986). We faced similar struggles trying to successfully clone an engineered plasmid, yet we could not identify the exact step that caused an unsuccessful engineering of the plasmid. Similar to other researchers, we may have faced struggles while trying to perform a restriction digest or a ligation. Although electroporation is more efficient and a critical step to the success of our project, the process can have random results that may have affected our out-comes. We were unable to successfully induce CRISPR to disrupt the *tnaA* gene in the HME63 strain of *E. coli*, but the future of CRISPR research is promising as it is already being used to edit other strains of bacteria, vertebrates, and even mammalian cells. Even though it is a fairly new technology, CRISPR's possibilities are endless as it is already being used to do amazing things in the biotechnology field.

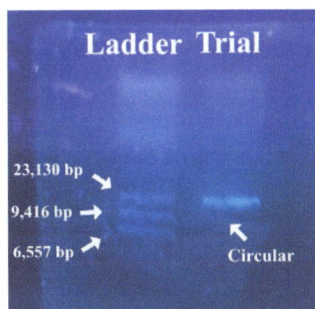

Picture 7: The above image shows a gel electrophoresis after an unsuccessful restriction digest. One trial was run through the gel alongside a λ DNA-HindIII Digest Ladder from NEB. Since only circular (non-cut) plasmids were visible, it indicates that restriction digest was unsuccessful.

ACKNOWLEDGMENTS

We would like to thank all those who have helped us in designing and conducting this research project. We would especially like to thank Dr. Megan Hochstrasser, the current Science Communications Manager at the Innovative Genomics Institute, for helping us throughout the design and execution of our project. She reviewed our methods and helped us troubleshoot problems we encountered throughout this research. We would also like to give a special thanks to Dr. Fan Yeung, a biology instructor at Front Range Community College, for allowing us to come to her lab to learn how to perform a miniprep along with helping us work through many of our other protocols, and Robert Kos, the technical sales and distribution sales representative at Harvard Bioscience, for helping us receive the electroporation machine and teaching us how to use it.

Another thank you to Susanne Petri, biotechnology instructor, for helping us in all aspects of our research. Thank you to Nayan Naik and the RCHS Science Department for their generous grant towards the purchase of materials needed to conduct our research, QIAGEN for donating the miniprep kit, Zymo Research for donating our gel purification kit, New England Biolabs for donating the restriction digest and ligation kits, and Recombineering for donating the HME63 bacteria. We would also like to thank previous research students Jonathan Ferry and Peter Deckerman for designing the gRNA and helping us learn how to perform protocols necessary for this research; Molly Dolan and Lauren McCaffrey for helping us learn how to perform protocols; and Megan Hupka and Madison Ditmer helping us perfect our research pitch before it was approved. We would also like to thank the following RCHS staff members: Bryan Winkelman, Teacher Librarian, for his hard work in designing and running our website, grading and editing our blog posts, helping us create a poster for presenting at science fairs, and all his efforts in guiding us through this project; and David Ferguson, chemical safety manager, for helping us get all of our chemicals and reagents approved and ensuring our safety through this research. We would also like to thank Rock Canyon High School and Douglas County School District for supporting the RCHS Biotechnology Program which has not only provided us with the necessary materials and lab space, but also provided us the opportunity to conduct authentic scientific research along with providing us a grant to buy a new electroporation machine that was necessary for our research.

REFERENCES

Addgene (2018). CRISPR History and Development for Genome Engineering. Retrieved from https://www.addgene.org/crispr/history/

Anslem , V. (2015). CRISPR-Cas9 mode of action. Retrieved from https://commons.wikimedia.org/w/index.php?title=File:CRISPRCas9_mode_of_action.png&oldid=241216510

Barrangou, R., & Horvath, P. (2010). CRISPR/Cas, the Immune System of Bacteria and Archaea. *Science. 327,* 167-170. doi:10.1126/science.1178555

Bhagavan, N.V., & Ha, C.-E. (2015). Tryptophan Operon. In *Essentials of Medical Biochemistry* (Regulation of Gene Expression). Retrieved from https://www.science direct.com/topics/biochemistry-genetics-and-molecular-biology/trp-operon

Center for Disease Control. (2005). *Staphylococcus aureus* biofilm. Retrieved from https://commons.wikimedia.org/wiki/File:Staphylococcus_aureus_biofilm_01.jpg

Cohen, J. (2017, February 15). How the Battle Lines Over CRISPR Were Drawn. Retrieved from http://www.sciencemag.org/news/2017/02/how-battle-lines-over-crispr-were-drawn

Cyranoski, D. (2016). First Trial of CRISPR in People. *Nature Journal, 535,* 476. Retrieved from https://www.nature.com/polopoly_fs/1.20302!/menu/main/topColumns/topLeftColumn/pdf/nature.2016.20302.pdf

Growth of Bacterial Cultures. (n.d.). Retrieved from https://www.qiagen.com/ch/resources/technologies/plasmid-resource-center/growth%20of%20bacterial%20cultures?akamai-feo=off

Ishino, Y., Shinagawa, H., Makino, K., Amemura, M., & Nakata, A. (1987). Nucleotide sequence of the iap gene, responsible for alkaline phosphatase isozyme conversion in Escherichia coli, and identification of the gene product. *Journal of Bacteriology. 169*(12), 5429-5433. doi:10.1128/jb.169.12.5429-5433.1987

Jamal, M., Ahmad, W., Andleeb, S., Jalil, F., Imran, M., Nawaz MA., ...Kamil, MA. (2018). Bacterial biofilm and associated infections. *US National Library of Medicine.* doi: 10.1016/j.jcma.2017.07.012

Jao, L., Wente. S. R., & Chen. W. (2013). Efficient multiplex biallelic zebrafish genome editing using a CRISPR nuclease system. *PNAS. 110*(34), 13904-13909. doi:10.1073/pnas.1308335110

Krishnavedala. (2014). Indole Synthesis. Retrieved from https://en.wikipedia.org/wiki/File:Indole.svg

Lee, J.H., & Lee, J. (2010) Indole as an intercellular signal in microbial communities. *FEMS Microbiology Reviews. 34*(4), 426-444. doi: 10.1111/j.1574-6976.2009.00204

Li, G., & Young, K. D. (2013). Indole production by the tryptophanase TnaA in *Escherichia coli* is determined by the amount of exogenous tryptophan. *Microbiology. 159*, 402-410. doi: 10.1099/mic.0.064139-0

MacWilliams, M. P. (2009). Indole Test Protocol. Retrieved from http://www.asmscience.org/content/education/protocol/protocol.3202

NEBcloner. (n.d.). Retrieved from http://nebcloner.neb.com/#!/

Ran, F. A., Hsu, P. D., Lin, C.-Y., Gootenberg, J. S., Konermann, S., Trevino, A. E., ...Zhang, F. (2013). Double Nicking by RNA-Guided CRISPR Cas9 for Enhanced Genome Editing Specificity. *Cell Press. 154*(6),1380-1389. doi: 10.1016/j. cell.2013.08.021

Riggs, C. D., & Bates, G. W., (1986). Stable transformation of tobacco by electroporation: Evidence for plasmid concatenation. *Proceedings of the National Academy of Sciences of the United States of America. 83*(15), 5602-5606. doi:10.1073/pnas.83.15.5602

Venkatachalam, V., Broadbent, K., Green, A., Payne, S., Zhu, B., Endy, D., ...Knight, T. (2006). Eau d'e Coli iGem 2006: MIT. Retrieved from https://2006.igem.org/wiki/index.php/MIT_2006

ABOUT THE AUTHORS

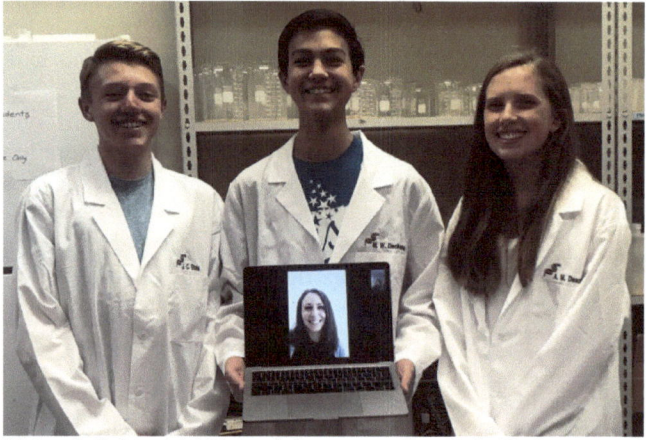

Pictured: (left to right) Pictured above are students Stohs, Deckerman, and Deschane with their mentor Dr. Megan Hochstrasser—the Science Communications Manager at Innovative Genomics Institute.

While CRISPR research brought many ups and downs, Deckerman, Deschane, and Stohs have gained many valuable skills including time management, communication, and public speaking. The protocols they used such as miniprep, restriction digest, gel electrophoresis, gel purification, ligation, and electroporation seemed impossible at first, but began to work after learning from their many failed trials. Although the project was unable to be brought to a conclusion, the course and research was incredibly beneficial and will have a lasting impact that stays with the researchers for the rest of their lives.

Inspired by his participation in the Rock Canyon High School Biotechnology Program and his older brother's previous research work, Deckerman, a junior at Rock Canyon High School, has developed a lot of enthusiasm in the field of biology. Conducting research at Rock Canyon High School has only increased his interest in the field of biology and allowed him to acquire excitement in conducting

his own research later in life. After high school, Deckerman intends to pursue a degree in microbiology to further fuel his passion for biological research.

After participating in the Introduction to Biotechnology course, Deschane knew she wanted to continue researching in the biotechnology field. After looking into many project ideas, CRISPR and its unlimited potential caught her eye and led to a year of research in the field. She has enjoyed conducting CRISPR and *E. coli* research in the Experimental Design in Biotechnology course and is now interested in pursuing these topics even further. After high school, Deschane plans on applying to a microbiology and biotechnology program, hopefully leading to a career in biotechnology.

Stohs has always had a love for science and has enjoyed the process of working with CRISPR his junior year at Rock Canyon High School. The research has made an incredible impact on his work ethic and given him an immersion into the biotech industry. He has been able to develop and refine a multitude of skills required for microbiological research and has been challenged in a way that has given him confidence in his problem solving abilities. After high school, Stohs plans to pursue a career in the research industry and continue learning new things about STEM and himself.

Effects of exposure to sanicle essential oil on the cellular proliferation of squamous skin cell carcinoma

C. J. Brand, A. J. Sonin, G. A. Hayrynen, & S. L. Fordham
Department of Science, Principles of Experimental Design in Biotechnology, Rock Canyon High School, Highlands Ranch, Colorado, USA

Essential oils are a fast-growing area of research due to anecdotal claims about their anti-cancer properties. One type of essential oil, sanicle, contains rosmarinic acid. This acid has been suggested in research to reduce the proliferation of cancer; however, sanicle essential oil is not scientifically backed or FDA-approved for cancer treatment. This research investigated the effects of sanicle essential oil on the proliferation of cancerous mouse squamous skin cells (A223) and non-cancerous mouse embryonic fibroblast cells (NIH3T3) *in vitro*. Cell counts were performed once a day for five days, and then analyzed with a non-parametric paired t-test, visually illustrated by regression models of proliferation, and the rate of proliferation compared. It was hypothesized that exposure to sanicle essential oil would lower the rate of proliferation of the squamous skin cell carcinoma (A223) cells due to the presence of rosmarinic acid which was supported as the data exhibited a strong trend of sanicle essential oil lowering the proliferation rate of squamous skin cell carcinoma (A223) cells. A223 cultures treated with sanicle essential oil had a 161% average proliferation rate compared to the non-treated control with a 270% average proliferation rate, suggesting that sanicle essential oil may be an effective treatment for squamous skin cell carcinoma. NIH3T3 cultures treated with sanicle essential oil had an average proliferation rate of 40% compared to the non-treated control with a 35% average proliferation rate, suggesting sanicle essential oil may have no significant effect on the proliferation rate of the non-cancerous mouse embryonic fibroblast cells (NIH3T3).

Three million cases of nonmelanoma skin cancer are diagnosed per year in the U.S. alone; of those, 700,000 cases are identified as squamous skin cell carcinoma (Cancer.net Editorial Board, 2012). Squamous skin cell carcinoma is most often caused by overexposure to ultraviolet light, although it can additionally be caused by ionizing radiation and human papillomavirus infection (Hawrot, Alam, & Ratner, 2003). The development of this cancer can be characterized by three main stages. Stage one, called initiation, is where normal squamous cells mutate due to exposure to carcinogens. Stage two, called pomotion, is where these mutated cells become inflamed and expand. Stage three, called progression, is when these mutated cells begin clonal proliferation and a tumor is formed. With detection in early stage three or earlier, this type of cancer is very treatable, however, when left undiagnosed, it can metastasize and become fatal. Squamous cells are found in the epithelial tissues that line various organs as well as in the outermost epidermal layer of the skin (**Fig. 1**). Current treatments include surgery, freezing, chemotherapy, or radiation, depending on the stage and location of the cancer (Bath-Hextall, Bong, Perkins, & Williams, 2004). These treatments have harsh and adverse side effects, including nausea, hair loss, anemia, and increased risk of infection (Coates *et al.*, 1983). Because of this, many people are beginning to seek alternative treatments and prevention methods, including the use of essential oils.

Essential oils are phyto products that that play an important role in the protection of their host plant from herbivores, insects, parasites, and fungus. Many essential oils exhibit bioactive properties, such as antimicrobial activity, and have been used throughout history for treating various diseases and ailments.

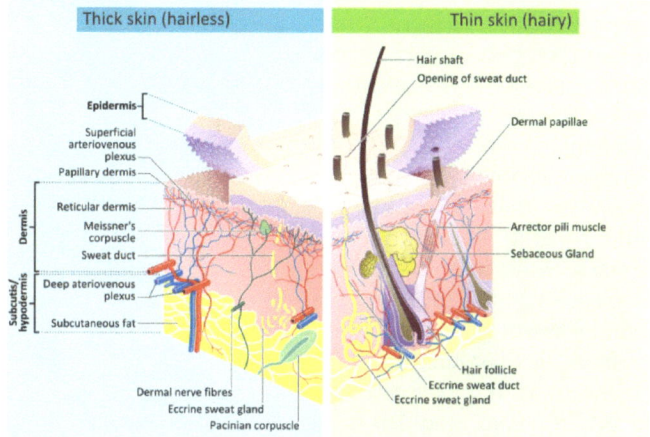

Figure 1: 3-dimensional model of human skin. Squamous skin cells are located in the epidermis, the uppermost layer. From "File:Skin Layers.svg" by Madhero88 and M. Komorniczak, 2012 (https://commons.wikimedia-.org/wiki/File:Skin_layers.svg). Copyright 2012. CC BY-SA 3.0

The highly active chemicals within essential oils, terpenes, consist of many different chemical structures all of which have a volatile nature (Blowman, Magalhães, Lemos, Cabral, & Pires, 2018). Approximately 3,000 different essential oils are known to exist, but only one-tenth have been scientifically tested for cosmetic or medical uses (Sharifi-Rad *et al.*, 2017). This means there are still many essential oils and their chemical constituents that have not been scientifically tested for their biomedical properties. Through documented research, essential oils have been shown to not only act as a preventative to cancer, but also to act directly on the tumor itself, having anti-proliferative properties, as well as apoptotic properties (Blowman *et al.*, 2018). In addition, research suggests that essential oils containing antioxidants may also reduce cancerous proliferation (Gautam, Mantha & Mittal, 2014). Because of this, essential oils have been marketed as an inexpensive, natural remedy for several ailments including cancer. If essential oils are marketed as a cosmetic, the seller's claims do not need to be approved by the FDA. However, if they are marketed as a drug, they must be approved. In either case, the contents and labeling of essential oils must be accurate under the Federal Food, Drug, and Cosmetic Act (FD&C Act) and the Fair Packaging and Labeling Act (FPLA) (U.S. Food and Drug Administration, 2000). Recently, however, multiple essential oil companies have been caught making medical claims about their products without FDA approval. In 2014, doTERRA, a multi-level marketing company, was sent warning letters by the FDA for claiming that their products can prevent diseases such as Ebola and cure cancer without the FDA's approval (U.S. Food and Drug Administration, 2014). Due to false or misleading claims by essential oil companies, skeptics have called into question the efficacy of essential oils for treating medical ailments. Because of the lack of scientific research to support the use of essential oils, more research still needs to be done on many essential oils and their components to better understand their potential anti-cancer properties. One such oil is sanicle essential oil.

Sanicle essential oil is extract-ed from *Sanicula europaea*, a perennial plant of the family *Apiaceae* (**Pic. 1**). This herbal treatment is most commonly used to treat coughs and bronchitis (WebMD, 2018). Companies that sell sanicle oil, however, claim it can improve artery health, decrease hemor-rhages, and prevent diseases such as syphilis or gonorrhea; and even cure cancer. Advocates of sanicle oil go as far as to claim it may cure "any bodily problem" and that it is a "tumor dissolver" when applied

Picture 1: *Sanicula europaea* being grown in the wild Sanicle essential oil has been frequently touted to possess multiple curative abilities, such as the ability to "dissolve tumors". From "File: Sanicle at Spier's. JPG" by Rosser1954, 2010 (https://commons.wikimedia.or g/wiki/File:Sanicle_at_Spier%2 7s.JPG). Public domain.

topically (Healed People, 2011). While these claims are neither backed by the FDA nor scientifically-based, there may be specific properties of sanicle oil that do in fact have positive health benefits (Talag, 2016). Sanicle essential oil contains four active chemicals: palmitic acid, bis(2-ethylhexyl) phthalate, saniculoside N, and rosmarinic acid. Rosmarinic acid, a component of sanicle essential oil, has antioxidant properties that have been shown in recent studies to reduce the incidence and proliferation of cancer and to decrease tumor size (Adomako-Bonsu, Chan, Pratten & Fry, 2017). Rosmarinic acid has also been found to reduce cyclooxygenase-2 and activator protein-1 expression in human cancer cells, which in turn reduces proliferation (Scheckel, Degner, & Romagnolo, 2008). Due to the absence of scientific research surrounding sanicle essential oil and cancer, we conducted an investigation into the anti-cancer potential of sanicle essential oil. We hypothesized that exposure to sanicle essential oil would reduce the proliferation rate of the A223 mouse squamous skin cell carcinoma line due to the presence of rosmarinic acid.

This investigation tested the effects of exposure to sanicle oil on the cellular proliferation of squamous skin cell carcinoma cells (A223) and non-cancerous embryonic fibroblasts (NIH3T3). This research did not aim to determine the molecular basis of any effects observed; instead, it focused on determining if the rate of proliferation in cancer cells was affected when compared to a non-cancerous control. In addition, cell death was not recorded. This research contributes to the growing scientific body of knowledge involving essential oils as potential treatments for cancer.

METHODS

In this research, the effects of exposure to sanicle essential oil on the cellular proliferation of A223 mouse squamous skin cell carcinoma cells was tested and compared to the effects on a non-cancerous NIH3T3 mouse embryonic fibroblast cell line to allow for a comparison between the effects of sanicle on the proliferation rate of cancerous and non-cancerous cells. Cell death was not documented; only the growth rate of the cells was recorded. Both cell lines were donated for this research by our mentor, Emily Duncan, with the University of Colorado Anschutz Medical Campus in the Department of Cell and Developmental Biology and are rated for work in a BSL-1 laboratory.

Experimental Design

Each of the two trials for each cell line spanned eight days with data collected on days four through eight. In addition to the treatment group, which consisted of Dimethyl Sulphoxide (DMSO) + sanicle oil, there were two controls for each cell line: DMSO + glycerol and cell culture media only (no treatment) (**Fig. 2**).

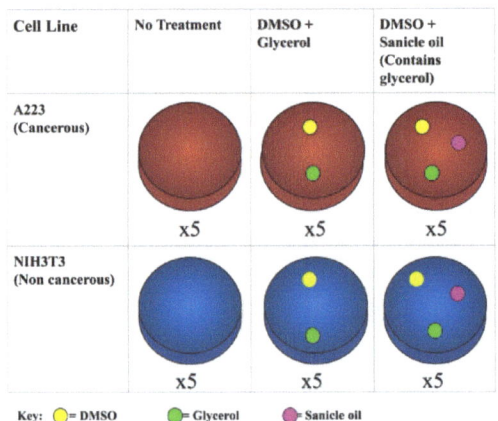

Figure 2: This table shows the experimental and two control groups that were used for each cell line with dots representing the treatments added to each culture. Each group consisted of five plates, one for each day that cell counts were performed. This was repeated for a total of two trials.

DMSO was used as an emulsifier to allow the sanicle to be absorbed into the cells. Due to the fact that sanicle oil is suspended in glycerol, glycerol was tested as a control alongside DMSO to determine any effects either chemical had on the proliferation of the cells. Five cell cultures of each treatment and control were prepared, one for each day that data was collected. After one culture was counted from each treatment or control each day, it was immediately discarded. The cell culture media consisted of 445 mL of DMEM (Dulbecco's Modified Eagle Medium), 50 mL Fetalgro Bovine Growth Serum (FBS), and 5 mL of 10,000 units/mL pen/strep stored at 4°C until use. To this stock media, we added the treatments: the sanicle + DMSO group was treated with 75 μL of sanicle and 20.4 μL of DMSO, while the glycerol + DMSO control group was treated with 2.5 μL of glycerol and 20.1 μL of DMSO, resulting in a constant 0.5% concentration of DMSO in each treated culture. Cell media was replaced every day on days four through eight, and the treatment media was added to the media starting on day four **(Pic. 2)**. Media changes were performed by aspirating the media, putting the same amount of stock media back into the dishes, and then adding appropriate treatments. Cells were incubated in a Sanyo MCO-17AC CO_2 incubator at 5% CO_2 and 37°C. NIH3T3 and A223 cells with a passage number of 5 were stored in liquid nitrogen, and 20 μL of cell solution was plated in each culture dish at the start of the trial.

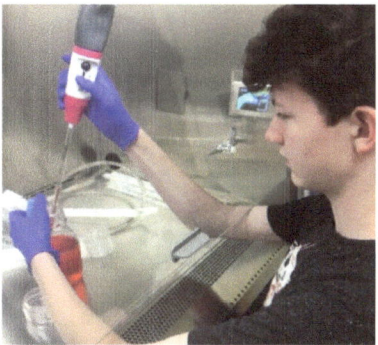

Picture 2: This image shows Sonin changing the cell culture media for the A223 cell cultures in the biological safety cabinet.

Plating Cells

Prior to experimentation, 1 mL of frozen cell solution was removed from liquid nitrogen storage, thawed, and plated into a 10 cm cell treat culture dish along with 8 mL of the stock media. They were then placed into the CO_2 incubator at 37°C with 5% CO_2 for one day. On day two, cells were split from the initial 10 cm culture dish into fifteen 6 cm culture dishes for the trial. 20 μL of the suspended cell mixture along with 4 mL of stock media was put into each 6 cm dish. On days four through eight, the stock media was aspirated and the media containing the treatment was added.

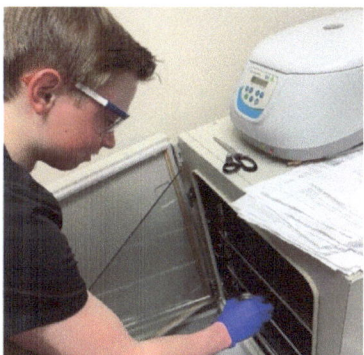

Picture 3: In this picture, Hayrynen is placing a cell culture dish into the CO_2 incubator during a trial with A223 cells. The CO_2 incubator provides the optimal environment for the cells to grow in.

Splitting Cells

When splitting the cells the protocols followed were consistent. To split the cells, the media was aspirated from the culture dish and the remaining adhered cells were rinsed twice with 1X PBS. Next, the cells were incubated with 0.5 mL of 0.25% trypsin at 37°C in the CO_2 incubator for 2-5 minutes (2 minutes for NIH3T3 and 5 minutes for A223) **(Pic. 3)**.

After incubation, 1.5 mL of the stock media was added to the cells to deactivate the trypsin. Next, the cell suspension was transferred into sterile 15 mL centrifuge tubes and centrifuged at 2500 rpm for 5 minutes. The supernatant was aspirated and the cell pellet resuspended with 4 mL stock media. When splitting from a 10 cm dish at the beginning of a trial, 20 μL of the suspended cell solution was transferred to a new 6 cm cell treat culture dish along with 4 mL of stock media.

Cell Counts

Cell counts were performed for each treatment and control over a five day period on days four through eight in the afternoon. Each day, one of the five culture dishes from each experimental and control group was counted and discarded. To count the cells, the same protocol for splitting cells was followed, however, after centrifuging and aspirating the supernatant, the cell pellet was resuspended and titrated with 4 mL of media. Then, 100 μL of the cell suspension was stained with 400 μL of 0.4% trypan blue and the live cells were counted using a hemocytometer. To visualize the cells, an EVOS FL inverted microscope with a 10x/0.30 objective on the bright light field setting was used **(Pic. 4)**. Then, after the count was taken, the amount counted was multiplied by $5 \cdot 10^4$ in order to determine the amount of viable cells per milliliter in the cell suspension.

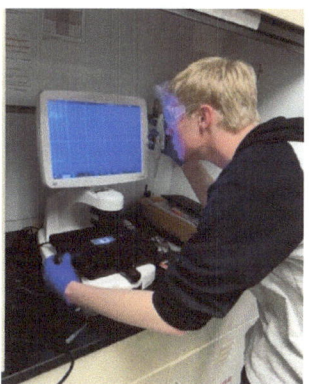

Picture 4: This image shows Brand using the EVOS FL microscope and the hemocytometer to perform a cell count of an A223 control plate on day four. The number of cells counted within a certain quadrant of the grid are inserted into an equation that gives the total cell count within the culture dish.

Data Analysis

For each trial, the average proliferation rate was calculated for the control groups and the experimental group. This rate was calculated by first calculating the percent change in cell count between each day of data collection on days four through eight. These four percentages were then averaged in order to obtain an average proliferation rate for each control and experimental group in the trial. A two-tailed paired t-test was used to calculate statistical significance between the average proliferation rates of the no treatment control to the experimental group as well as the no treatment control to the glycerol control groups.

RESULTS

In this research, the effects of sanicle essential oil on the proliferation rate of cancerous A223 mouse squamous skin carcinoma cells were compared to a non-cancerous cell line, NIH3T3 mouse embryonic fibroblast cells. The cell cultures were grown *in vitro*. Each trial consisted of one treatment and two controls for each cell line. The proliferation rate of the A223 squamous skin cell carcinoma cultures treated with sanicle was lower than the no treatment and glycerol only controls, with the average proliferation rate in the sanicle-treated cells at 161% compared to the no treatment control at 270% (**Graph 1**). In comparison, the cell count throughout the experiment as well as the average proliferation rate of the glycerol only control closely matched the no treatment control (**Graph 2**). This suggests the reduction in cellular proliferation in the sanicle-treated cells was due to the sanicle oil and not the glycerol base. It is also noteworthy that the cell count between days seven and eight in the sanicle-treated A223 cell line actually decreased, which suggests that cell death/apoptosis was induced (**Graph 2**).

Trial one had a much higher average proliferation rate than what was observed in trial two for all groups due to a low cell count on the first day of the trial, which made the proliferation rate of the first to second days considerably higher than the rest of the days. When analyzed using a paired t-test, the difference in the rate of cellular proliferation between the sanicle treatment and the no treatment control in the A223 cell line resulted in sanicle-treated cells having a

lower average proliferation rate which was trending towards being statistically significant (p = 0.0498). We also found that there was no trend towards statistical significance between the no treatment control and glycerol control groups (p = 0.6223). Statistical significance could not be verified, however, due to the limited number of trials that were performed. No morphological changes were observed for any of the control groups or the experimental group, however, an abundance of dead cells were observed in the sanicle-treated cultures. While cell death was not directly measured, this observation supports the reduction in cell numbers between days seven and eight.

In addition to testing the effects of sanicle essential oil on mouse squamous skin cell carcinoma, we also tested its effects on a noncancerous cell line, NIH3T3 mouse embryonic fibroblasts. The effects of sanicle essential oil on the proliferation rate in this cell line were inconsistent between and throughout each trial. However, the average ending cell count of the no treatment control group was the highest at 1,575,000 cells/mL, while the average ending cell count of the sanicle treated group was significantly lower at 550,000 cells/mL (**Graph 3**). The average proliferation rate of the NIH3T3 cultures followed no particular trend, with the average proliferation rate of sanicle being the lowest in trial one at 35% and the highest in trial two at 45% (**Graph 4**). In addition, the control and glycerol had the same average proliferation rate at 43% in trial one, but significantly different proliferation rates in trial two, with control at 27%

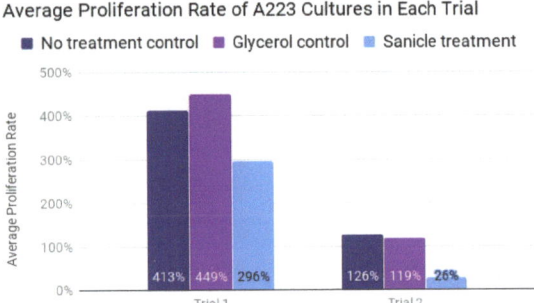

Graph 1: This graph shows the average proliferation rate of each culture in each of the A223 trials. In both trials, the average proliferation rate of the sanicle-treated experimental group was significantly less than the control groups, while the two control groups in each trial had similar average proliferation rates.

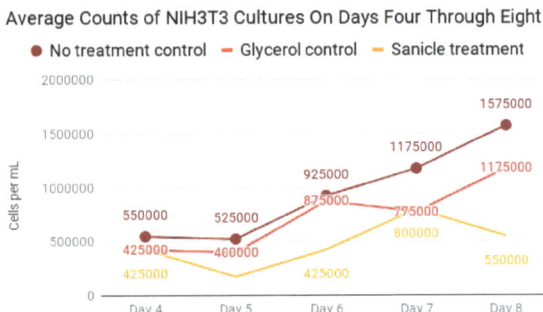

Graph 3: This graph shows the average counts of the NIH3T3 cultures over the course of the trials. The data here are very inconsistent, with counts varying inconsistently from day to day in all three groups. However, the final count of sanicle was the lowest, and the count of control was the highest at the end of the trial.

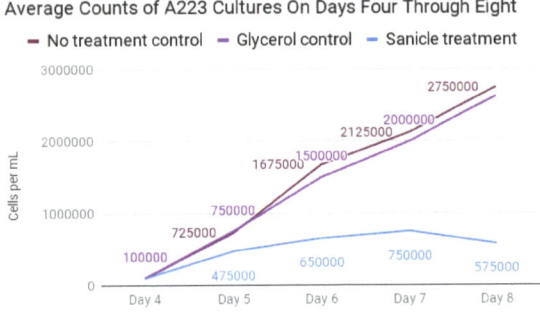

Graph 2: This graph shows the average cell counts for the A223 mouse squamous skin cell carcinoma cell line on days 4-8. The sanicle treatment group, shown in pink, demonstrated a flat and eventually negative growth rate. The two controls, no treatment and glycerol, both demonstrated a steadily increasing rate of proliferation.

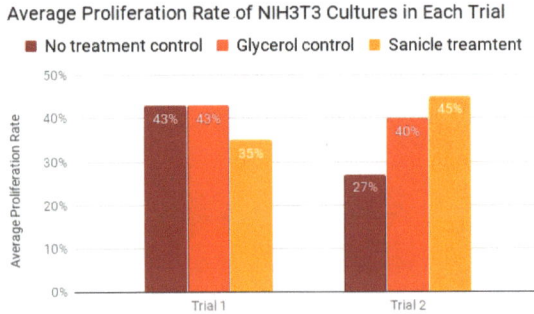

Graph 4: This graph shows the average proliferation rate of each group of each trial of the NIH3T3 cell line. The data here are inconsistent, with the sanicle treated group being the lowest in trial one, but the highest in trial 2. In addition, the control and glycerol groups had the same average proliferation rate in trial 1, but the control group had a significantly lower average proliferation rate than the glycerol treated group in trial 2.

and glycerol at 40%. No morphological changes were observed in either of the control groups or the experimental group, however, an abundance of dead cells was not noticed in this cell line as was observed for the A223 cell lines. This was not confirmed through direct measurements.

DISCUSSION

Sanicle essential oil has been anecdotally claimed to be a treatment for skin cancer, with claims even touting its ability to dissolve tumors. While these claims have not been confirmed through FDA backed studies, sanicle essential oil contains rosmarinic acid, which in prior research has found to reduce the proliferation rate of cancer cells (Adomako-Bonsu, Chan, Pratten, & Fry, 2017). However, sanicle essential oil and some of its other chemical constituents, such as saniculoside N, have not been scientifically researched for their anti-cancer properties. We tested the effects of sanicle essential oil on the proliferation rate of cancerous mouse squamous skin cells (A223) and noncancerous mouse embryonic fibroblast cells (NIH3T3) in order to test the validity of these claims. We hypothesized sanicle essential oil would reduce the proliferation rate of the A223 mouse squamous skin cell carcinoma cell line due to the presence of rosmarinic acid.

The A223 mouse squamous skin cell carcinoma cultures treated with sanicle essential oil demonstrated a lower proliferation rate when compared to the no treatment control, with average proliferation rates of 161% and 270%, respectively **(Graph 1)**. The control cultures proliferated at steady rates throughout the trials, while the sanicle treated culture's growth slowed down and even decreased in number at the end of the trial **(Graph 2)**. While the p-value (p = 0.0498) demonstrated statistical significance, suggesting that sanicle essential oil might have the ability to reduce the proliferation rate in mouse squamous skin cell carcinoma, not enough trials were performed for statistical significance to be accurately determined. The glycerol control, in contrast, did not exhibit a significantly different proliferation rate when compared to the untreated control group (p = 0.6223), suggesting that the glycerol base used in the sanicle essential oil had no effect on the proliferation rate of the cells. It was also observed that the cultures containing A223 cells treated with sanicle essential oil had large numbers of dead cells floating in the cell media. This was evident specifically in the A223 data when the cell counts decreased between days 7 and 8 at the end of the trial, with average cell counts decreasing from 750,000 to 575,000 cells/mL. While we did not directly count or measure cell death in this experiment, this is a noteworthy finding that warrants further investigation, as this may mean that exposure to sanicle essential oil not only has the ability to lower the cellular proliferation rate of squamous skin cell carcinoma, but also may have apoptotic effects.

In addition to testing the effects of sanicle essential oil on the cancerous A223 cell line, we also used a noncancerous cell type to determine if any effects observed were specific to cancerous cells. Neither the NIH3T3 mouse embryonic fibroblast cultures treated with glycerol nor those treated with sanicle essential oil had a significantly different proliferation rate when compared to the no treatment control (p = 0.5192 and p = 0.7778 respectively), suggesting that neither the sanicle nor the glycerol had an effect on the proliferation rate of noncancerous mouse embryonic fibroblast cells. It is important to note that the NIH3T3 cell counts were very inconsistent, with the average counts for the glycerol and sanicle treated cultures growing at erratic rates throughout the trials. For example, the count of the glycerol culture went from 400,000 on day five up to 875,000 on day six, down to 775,000 on day seven, and back up to 1,175,000 on day eight **(Graph 3)**. The average proliferation rate of the NIH3T3 cultures were also inconsistent from trial to trial, with sanicle treated cultures being the lowest in trial one at 35%, and sanicle treated cultures being the highest in trial two at 45% **(Graph 4)**. In addition, in the sanicle treated cultures, we did not observe a high number of dead cells in the media like there were in the cancerous A223 cell cultures **(Pic. 5)**. Although these data were not quantified, this might suggest that sanicle essential oil is capable of inducing apoptosis in A223 mouse squamous skin cell carcinoma, but not NIH3T3 mouse embryonic fibroblast cells. Similar to the A223 trials, not enough trials of the NIH3T3 cells were performed to determine statistical significance of the data, but it is interesting to note that the proliferation rate of the sanicle treated NIH3T3 cells were much closer to their corresponding controls when compared to the A223 cell line. This suggests that sanicle may have properties that are not only inhibiting cell division and growth, but may also be specific to mouse squamous skin cell carcinoma.

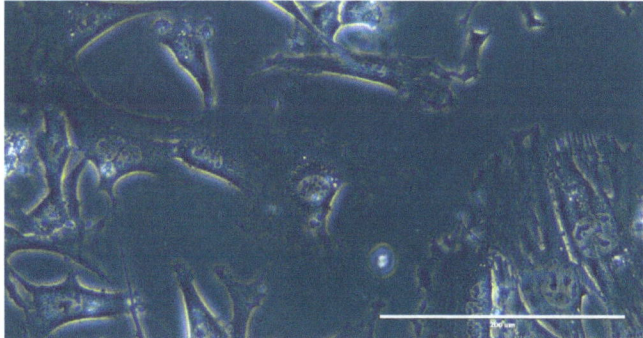

Picture 5: This picture shows a healthy NIH3T3 cell culture treated with sanicle with no abundance of dead cells.

This study, while an important first step in understanding the effects of sanicle oil on skin cancer, has several limitations. First, we were unable to quantify the exact amount of sanicle oil that had been absorbed by the cells. With the addition of DMSO to the culture media, we were confident we had provided a means for the cells to absorb the sanicle oil; however, the exact amount that was absorbed remains unknown. In addition, only five cell counts were performed for each experimental group. After analyzing the data, we feel that having had a longer period of time for the cells to grow and to perform cell counts would have been more informative. In addition, even though we observed high rates of cell death within the sanicle-treated A223 cells, accurate counts of dead cells were not taken. Therefore, we cannot accurately conclude if the treatments killed cells or induced apoptosis in either cell line. Due to time limitations, only two trials were conducted for each cell line, which typically is not considered enough data to accurately report statistical significance. Additionally, the inconsistency of the NIH3T3 trials makes it difficult to accurately report statistical significance. As a result, additional trials must be conducted for both cell lines. However, if these strong trends continue in future trials, then it could provide further evidence to support that sanicle essential oil may be an effective treatment for slowing the proliferation rate of squamous skin cell carcinoma. Because most conventional treatments for cancer, such as surgery, freezing, and chemotherapy are invasive and can result in adverse side effects, a less invasive aid to these treatments would be beneficial (Coates *et al.*, 1983). If sanicle essential oil reduces proliferation in squamous skin cell carcinoma, it could possibly be used alongside or as an alternative to conventional treatments to improve success rate and reduce the period of treatment.

Once more trials are performed, there are several important next steps that should be taken to further the knowledge of this essential oil. The first would be to quantify the amount of apoptotic cells in the cultures to determine if sanicle induces apoptosis or merely reduces the proliferation rate **(Pic. 6)**. Traditional cancer treatments, such as chemo-therapy, induce apoptosis in dividing cells, and they do so in cancerous cells as well as various other healthy cells in the body by inducing the excretion of cytochrome c from the mitochondria and the ligation of cell surface death receptors (Kaufmann & Earnshaw, 2000). If sanicle essential oil is able to target cancer cells specifically and induce apoptosis, then it could reduce tumors without affecting healthy cells. The purpose of this study was not to determine the mechanism of action and the chemical/metabolic pathways sanicle essential oil is impacting, however, it is an important next step. Sanicle essential oil contains rosmarinic acid, and rosmarinic acid has been shown to suppress activator protein-1-dependent activation of the COX-2 enzyme, which is responsible for promoting tumor growth (Scheckel, Degner, & Romagnolo, 2008). However, because the additional components in sanicle essential oil have not been tested on the proliferation rate of cancer, this effect may not be solely due to the presence of rosmarinic acid, or at all. This information would be useful for isolating the chemical compound(s) within sanicle essential oil that possess the desired anti-cancer property and could inform future medical

practices. It would also be useful to quantify the amount of sanicle oil absorbed by the cells to determine the amount of oil required for possible treatments. Testing sanicle essential oil on other cancerous and noncancerous cell lines would also be useful to determine if sanicle is effective for treating different types of cancer and if it is safe to use on different areas of the body. Additionally, testing sanicle oil *in vivo* is necessary to determine if the effects of sanicle can be reproduced *in vivo*, and the optimal method of delivery. Overall, if the strong trends observed in this research continue, sanicle essential oil may be an effective cancer treatment.

ACKNOWLEDGMENTS
First, we would like to thank Emily Duncan, a PhD student at the University of Colorado Anschutz Medical Campus in the Department of Cell and Developmental Biology. She mentored us throughout the research, helped organize methods, and provided the cell lines and some cell culture materials. We would also like to thank Madison Dittmer and Kylie Hutchison, seniors at Rock Canyon High School who performed cell culture in the Biotechnology Research Program last year. They trained us on how to perform cell culture, helped us with troubleshooting problems we encountered, and worked with us on our presentation skills. We would also like to thank Susanne Petri, our biotechnology instructor. She supported us throughout every aspect of this research project. We would like to thank Nayan Naik and the RCH Science Department for funding our research which allowed us to purchase the necessary materials needed for this research. Several RCHS staff members assisted us and supported our research this year. We greatly appreciate their time and efforts: chemical safety managers David Ferguson and Kerry Hinton for reviewing the safety of the materials we worked with in this experiment; Teacher Librarian Bryan Winkelman for assisting us with writing and editing our scientific paper, blog posts, posters, and webpage; and statistics teacher Gwen Karaba for assisting us with statistics. Finally, we would like to thank Rock Canyon High School and the Douglas County School District and its teachers for their support of the Biotechnology Program and for providing the laboratory space and the education that allowed us to perform our research.

REFERENCES
Adomako-Bonsu, A. G., Chan, S. L., Pratten, M., & Fry, J. R. (2017). Antioxidant activity of rosmarinic acid and its principal metabolites in chemical and cellular systems: Importance of physico-chemical characteristics. *Toxicology in Vitro*, *40*:248-255. doi:10.1016/j.tiv. 2017.01.016

Bath-Hextall, F., Bong, J., Perkins, W., & Williams, H. (2004). Interventions for basal cell carcinoma of the skin: Systematic review. *BMJ*, *329*(7468), 705. doi:10.1136/bmj.38219.515266.ae

Blowman, K., Magalhães, M., Lemos, M. F., Cabral, C., & Pires, I. M. (2018). Anticancer Properties of Essential Oils and Other Natural Products. Evidence-Based Complementary and Alternative Medicine, 2018, 1-12. doi:10.1155/2018/3149362

Cancer.net Editorial Board. (2012, June 25). Skin Cancer (Non-Melanoma) - Statistics. Retrieved from https://www.cancer.net/cancer-types/skin-cancer-non- melanoma/statistics

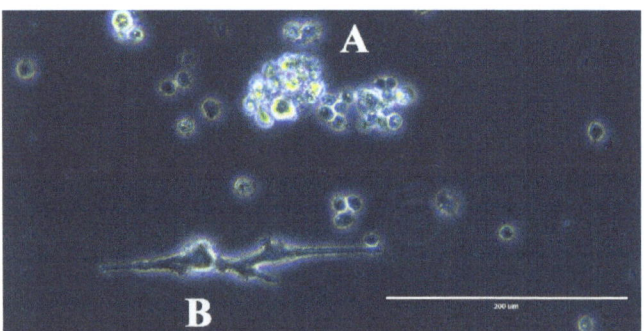

Picture 6: This picture shows a clump of apoptotic A223 cells with many other dead cells surrounding it (A) and two live cells (B). The apoptotic cells should be quantified in future research.

RESEARCH IN BIOTECHNOLOGY

Coates, A., Abraham, S., Kaye, S., Sowerbutts, T., Frewin, C., Fox, R., & Tattersall, M. (1983). On the receiving end—patient perception of the side-effects of cancer chemotherapy. *European Journal of Cancer and Clinical Oncology, 19*(2), 203-208. doi:10.1016/0277-5379(83)90418-2

Gautam, N., Mantha, A. K., & Mittal, S. (2014). Essential Oils and Their Constituents as Anticancer Agents: A Mechanistic View. BioMed Research International, 2014, 1-23. doi:10.1155/2014/154106

Hawrot, A., Alam, M., & Ratner, D. (2003). Squamous cell carcinoma. *Current Problems in Dermatology, 15*(3), 91-133. doi:10.1016 /s1040-0486(03)00005-x

Healed People. (2011). Natural Herbs To Dissolve Tumors. Retrieved from https://www.healedpeople.com/knowledge/natural-health-alternatives/ 41-pure-herbs/44-natural-herbs-to-dissolve-tumors

Kaufmann, S. H. & Earnshaw, W. C. (2000). Induction of Apoptosis by Cancer Chemotherapy. Experimental Cell Research, *256(*1), 42-49. doi:10.1006/excr.2000.4838

Komorniczak, M. (2012, October 9). Skin Layers - Wikimedia Commons. Retrieved from https://commons.wikimedia.org/wiki/File:Skin layers.svg

"Sanicle at Spier's. JPG" by Rosser1954, (2010) https://commons.wiki-media.org/wiki/File:Sanicle_at_Spier%27s.JP

Scheckel, K. A., Degner, S. C., & Romagnolo, D. F. (2008). Rosmarinic Acid Antagonizes Activator Protein-1–Dependent Activation of Cyclooxygenase-2 Expression in Human Cancer and Nonmalignant Cell Lines. *The Journal of Nutrition, 138*(11), 2098-2105. doi: 10.3945/jn.108.090431

Sharifi-Rad, J., Sureda, A., Tenore, G., Daglia, M., Sharifi-Rad, M., Valussi, M., . . . Iriti, M. (2017). Biological Activities of Essential Oils: From Plant Chemoecology to Traditional Healing Systems. Molecules, 22(1), 70. doi:10.3390/molecules22010070

Talag, A.H. (2016). phytochemical investigation and biological activities of Sanicula europaea and Teucrium davaeanum. Isolation and identification of some constituents of Sanicula europaea and Teucrium davaeanum and evaluation of the antioxidant activity of ethanolic extracts of both plants and cytotoxic activity of some isolated compounds (Doctoral dissertation). Retrieved from Bradford Scholars database. (Accession http://hdl.handle.net/10454/14482).

U.S. Food and Drug Administration. (2000, March 13). Aromatherapy. Retrieved from https://www.fda.gov/cosmetics/productsingredients/ products/ucm127054.htm

U.S. Food and Drug Administration. (2014, September 22). Warning Letter doTERRA. Retrieved from https://www.fda.gov/iceci/ enforcementactions/warningletters/2014/ucm415809.htm

WebMD. (2018). Sanicle: Uses, Side Effects, Interactions, Dosage, and Warning. Retrieved from https://www.webmd.com/vitamins/ai/ ingredientmono-86/Sanicle

ABOUT THE AUTHORS

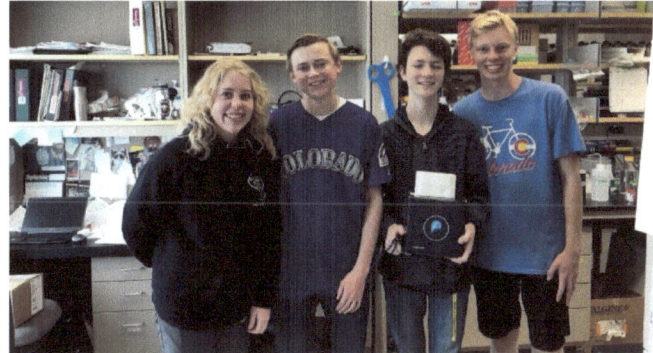

Pictured: Mentor Emily Duncan with Hayrynen, Sonin, and Brand at the CU Anschutz Medical Campus.

Brand, Sonin, and Hayrynen formed a team at the end of tenth grade. Together they developed many important skills including time management, presentation, communication, responsibility, and teamwork. Applying these skills and utilizing their great teamwork throughout the year led them to great accomplishments. They successfully completed their research project and got very interesting results. Their new lab and problem solving skills will benefit them for the rest of their lives.

Brand is a junior at Rock Canyon High School. He has a deep passion for science, especially in the field of medicine, where he has high ambitions for the future. In pursuit of this passion for medicine, he took the Experimental Design in Biotechnology and developed this project along with Sonin and Hayrynen. In the future he hopes to work as a trauma surgeon as well as a medical researcher. Aside from medicine, he has a deep interest in physics, specifically particle physics. He is not only an ambitious student, he is also an elite athlete. He is a nationally ranked triathlete and an avid cyclist. He hopes one day to ride in the Tour De France. Aside from his student-athlete life, he enjoys volunteering at the ICU at Sky Ridge Medical Center as well as hanging out with friends and journeying in the mountains.

Sonin is a junior at Rock Canyon High School. He became interested in biology when he took biology in 8th grade. He was very interested in Rock Canyon High School Biotechnology, so he took honors Biology freshman year, and Introduction to Biotechnology his sophomore year. This year he decided to further investigate research and continue his work in Experimental Design in Biotechnology. Along with Hayrynen and Brand, he is proud of the research he conducted this year. He also likes to code in his free time, and he's very interested in computer science as well. He likes to play video games with his friends, ski during the winter, and he likes to sleep.

Hayrynen is a junior at Rock Canyon High School. He became interested in the field of science when he was first exposed to Chemistry and the Introduction to Biotechnology class. He then got fully involved in the field of chemistry and biotechnology through the offerings at RCHS. He then partnered with two motivated individuals to conduct this research. Not only does Hayrynen dedicate himself to science, he is also actively involved in numerous clubs, weekly volunteering at the Spine and Total Joint Center at Sky Ridge Medical Center, and on the Youth Roots student board which takes action on the needs of his community. He is also involved in several sports and takes part in outdoor adventures, like camping and fishing, during his free time.

Danio rerio embryos as a model organism of secondary injury mechanisms in traumatic brain injury research

V. Elango, A. S. Kozlowski & S. M. Petri
Department of Science, Principles of Experimental Design in Biotechnology, Rock Canyon High School, Highlands Ranch, Colorado, USA

Traumatic brain injury (TBI) is an injury to the brain causing mild to severe damage that requires treatments such as rest, medications, and rehabilitation. In TBI, secondary injuries continue after the initial impact, affecting neurocognitive functions. TBI is a leading cause of death and disability in the United States, so it is important to further study the mechanisms of TBI and identify potential drug targets. Invertebrate model organisms, such as *Danio rerio* (zebrafish) embryos, are needed for early stage TBI research as they are useful as a pharmaceutical screen and target identification tool. Our research evaluated the ability of zebrafish embryos, until 6 days post fertilization (dpf), to serve as models of post-TBI secondary injuries and neurological symptoms. We used a shaking incubator at 250 rpm to induce head trauma in zebrafish embryos at 3 dpf, after which we evaluated the subsequent mortality, response to stimulus, and head size. The mortality assay demonstrated successful injury with an average of 11.7% mortality with treatment, within the 10-25% expected rate. The touch assay and head size assay showed little difference between the control and experimental groups and resulted in differences that were statistically insignificant across all trials. This data did not demonstrate evidence of successful induction of TBI in the embryos and it suggests that zebrafish embryos may not be a viable model organism for TBI research when using a shaking incubator to induce TBI. Future research is needed to determine if a different protocol would be more optimal.

Traumatic brain injuries (TBI) are one of the leading causes of disability and death in youth and young adults in the United States, with 2.8 million diagnosed cases annually and an average mortality rate of 50,000 people per year (Taylor, Bell, Breiding, & Xu, 2017). A TBI is defined as a hard hit to the head, causing the brain to move violently in the skull (Prins, Greco, Alexander, & Giza, 2013). The most common causes of TBI related injuries include falls and motor vehicle accidents (Taylor *et al.*, 2017). TBI leads to neurological symptoms including memory loss, sensitivity to light and sound, loss of consciousness, or headaches. One of the most common symptoms of TBI is a decrease of processing speed in the affected patient (P. Thompson, personal communication, August 22, 2018). Bruising, bleeding, and inflammation of the brain are just a few of the physical symptoms of TBI. There are two parts of a TBI, the primary and secondary injury. The primary injury is the initial force that causes the secondary injuries, which are defined by the sequence of events that occur after the primary injury. The events can be intra or extracranial, such as ischemia, hypotension, or edema (Murthy, Bhatia, Sandhu, Prabhakar, & Gogna, 2005). Neurological effects such as delayed response times, loss of basic cognitive functions and edema (the swelling and increase in intracranial pressure from fluid) are all secondary injuries that occur after a TBI (Unterberg, Stover, Kress, & Kiening, 2004). In a TBI, the neuronal membranes weaken, allowing for the release of free radicals into the extracellular space. These free radicals are different proteins and chemicals that commonly do not exist outside of the neuron, and can be tagged using biomarkers. In addition to brain dysfunction, scientists have noted visible damage to the axon of the neurons with TBI (Katz, Cohen,

& Alexander, 2015). The axon is the nerve fiber between the neuron and is used as communication between different neurons, muscles, and glands; Focal axonal swelling is a clear indicator of TBI that results in changes in cognition, memory, and behavior (Rudy, Maia, & Kutz, 2016; Johnson, Stewart, & Smith, 2013).

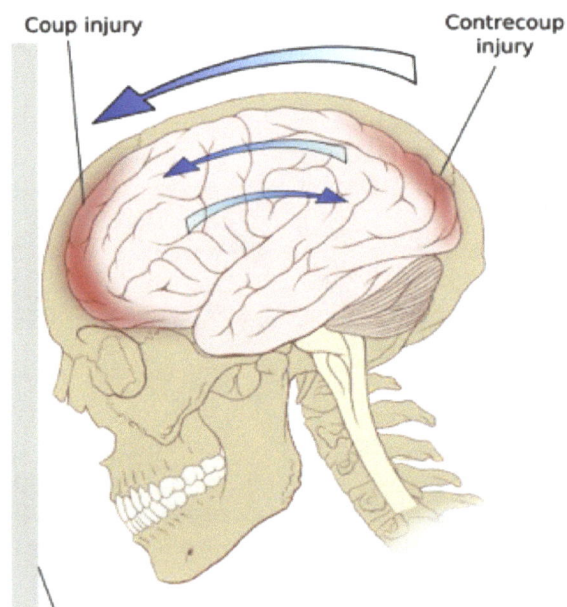

Figure 1: This figure shows one type of TBI, Coup-Contrecoup Injury where both ends of the brain is injured. From "Contrecoup" adapted from "Skull and brain normal human," by P. J. Lynch, 2006, https://commons.wikimedia.org/wiki/File:Skull_and_brain_normal_human.svg. Copyright 2006 by Creative Commons Attribution 2.5 License.

TBI is a general term for an injury to the brain that is caused by a force outside the body. The many types of TBI include Coup-Contrecoup (**Fig. 1**), Brain Contusion, Diffuse Axonal Injury, Second Impact Syndrome, Shaken Baby Syndrome, Penetrating Injury, and Concussion (Spinal Cord, 2018). A concussion is a mild form of TBI. The number of concussions is on the rise for all age groups. Every year, 1.6 to 3.8 million concussions are diagnosed in the United States alone. Between 2001-2005, children aged 5-18 accounted for approximately 135,000 of these concussions. Students in high school are increasingly becoming diagnosed with concussions. At Rock Canyon High School in 2017 alone, 79 students, many of whom are athletes, were diagnosed with a concussion which was 3.6% of the entire student body (G. Sims, personal communication, August 18, 2018). Per year, 10% of all athletes sustain a head injury serious enough to be considered a concussion (Protect the Brain, 2018). Concussions are more detrimental in children aged 0-14, whose brains are not yet fully developed, and adults 65 years and older, whose brains are unable to recover from injury as efficiently. Sustaining a concussion at such a young age when the brain is not fully developed can have drastic, lifelong consequences such as impaired cognitive function, chronic headaches, and for some, chronic traumatic encephalopathy (Bullock, 2018). For these reasons, further research into the mechanisms related to TBI and concussions to identify new treatments and drug targets is critical.

Model organisms of TBI and concussions allow researchers to gain insight into the mechanisms of both the primary and secondary injuries associated with TBI. Currently, research is being conducted on adult zebrafish as they are a popular vertebrate model. Understanding TBI mechanisms and developing treatments using zebrafish models is very important due to their neurological regenerative abilities (Cacialli, Palladino, & Lucini, 2018). Prior research conducted on the neurological effect of TBI in zebrafish adults using a stab lesion assay demonstrated the ability of the neuronal regeneration mechanisms to repair damaged tissues and neurons in the impacted area (Schmidt, Beil, Strähle, & Rastegar, 2014). In other research, a small weight dropped on the heads of adult zebrafish induced changes to the natural movement of the fish, indicative of TBI (Maheras *et al.*, 2018). Per the NIH, zebrafish are not considered vertebrates until 7 dpf and as such require less stringent regulation compared to other vertebrate models (National Institute of Health, 2016). Having a model of TBI in zebrafish embryos would enable scientists to quickly and more easily perform early stage drug screens as well as learn more about the cellular mechanisms involved with TBIs, ultimately leading to faster drug and treatment development. In addition to performing early stage drug screens, scientists would be able to analyze the developmental effects of TBI on zebrafish embryos which may be relatable to the brain development in children post TBI. Embryonic research is important to gain a diverse and comprehensive understanding of the mechanisms involved with TBI. For this research we used response to touch and head swelling as the secondary injuries used to identify if there was a TBI.

While adult zebrafish have served as a good model organism for novel TBI therapeutics, embryos have yet to be used for TBI research. Zebrafish embryos would serve as a useful model for initial drug screening assays due to their low cost, high progeny rate, small size, and lowered regulation as a vertebrate models (Chitramuthu, 2013). One adult breeding pair can produce up to 200 embryos per week, making them a good model organism for early stage research (Zebrafish Health, 2010). Up through 3 dpf, they are housed within a clear chorion that provides protection from the external environment. Zebrafish embryos are transparent, develop externally, and their development is well documented and

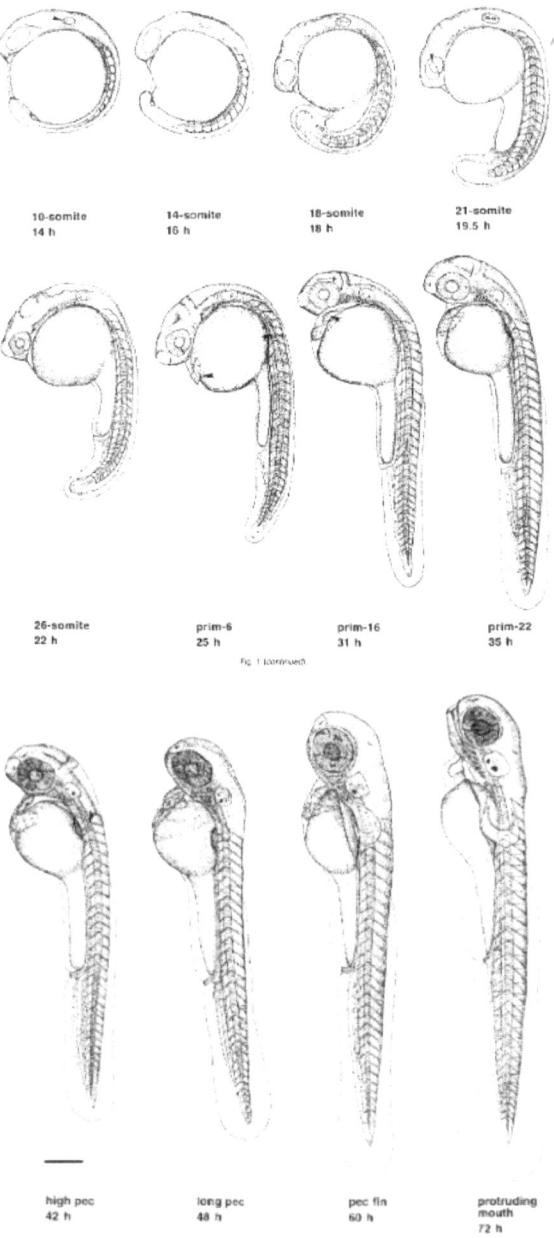

Figure 2: The progression of a zebrafish embryo throughout the major developmental stage through 3 dpf. Reprinted from "Stages of Embryonic Development of the Zebrafish," by C. Kimmel, W. Ballard, S. Kimmel, B. Ullmann, & T. Schilling, 1995, *Development Dynamics: an official publication of the American Association of Anatomist, 203*(3), p. 258-259. Copyright 2005 by John Wiley and Sons. Reprinted with permission.

characterized (**Fig. 2**) (Your Genome, 2014). Approximately 70% of zebrafish genes are homologous to humans and they have organs and systems similar to humans including the brain, which contain neurons similar to that of our own and similar brain developmental stages.

In our research, we engineered and tested a method to induce TBI in zebrafish embryos in order to determine if they can be a useful model organism for TBI research. We used a shaking incubator to induce head trauma and evaluated the subsequent mortality at several different rpms. Our goal was to identify a TBI inducing method that resulted in 10-25% mortality based on prior research (Prins *et al.*, 2013). Having some degree of mortality was necessary to ensure that the method was not only inducing harm but also was doing so at a level that is consistent with natural TBI. Based on these mortality assays, we determined that 250 rpm was the optimal setting for potentially inducing TBI. We used a shaking incubator due to its ease of use and accessibility in most laboratory settings. We conducted two experimental trials, using this method, and measured the effects on the subsequent response to stimulus and head size. Due to previous research in adult zebrafish, we hypothesized that TBI induced head trauma in zebrafish embryos would result in significant changes in these two assays making them an effective model organism for future TBI research.

METHODS
This research analyzed the ability of zebrafish embryos to serve as future model organisms for TBI related research. To be an effective model, the embryos must demonstrate clear measurable TBI symptoms following injury to the brain. The developing embryos were exposed to an external force designed to create trauma to the brain and the subsequent mortality rate, response to touch, and head size was evaluated.

Zebrafish Care and Breeding
In order to perform our research, we bred our own adult zebrafish and collected their embryos. To care for the adult zebrafish properly, we housed them in an in-home environment in two 10-gallon tanks; females were kept in one tank

Picture 1: This photo shows the Casper fish females in the 20 gallon tank. Females were in one tank and males were kept in an identical tank.

separate from the males (**Pic. 1**). The water was kept at a constant temperature of 28.5℃ and a pH of 6.8-7.5 in order to provide ideal conditions for breeding and development. The pH as well as levels of ammonia, nitrate, and nitrite were measured weekly using a colorimetric test (API Master Test Kit). Ammonia and nitrite levels were kept as close to zero as possible, and nitrate was maintained at less than 50 mg/L. The tanks were kept on a constant light cycle of 14 hours of light and 10 hours of darkness in an area with minimal disruption. The adults were fed a daily diet consisting of fish flakes, dry brine shrimp, and freeze dried bloodworms; protein enriched food was always fed to the fish the day before breeding. The night before breeding, one female and two males were placed into the breeding tank (Carolina Zebrafish Breeding Tank), separated by a plastic divider, and left in complete darkness overnight. In the morning, the lights were turned on at 7 am and the water inside the breeding tank was refreshed. The divider was removed allowing the males and females to be together to breed. After there were visible eggs in the bottom of the breeding tank, the adults were returned to their original tanks and the embryos strained, rinsed, and placed in a 10 cm petri plate containing 100 mL of egg water. The egg water consisted of a ratio of 40 g of Instant Ocean Aquarium Salts to 1 L of distilled water to make stock salts, and 1.5 mL of stock salts for every 1 L of distilled water (The Zebrafish Information Network, 2013). The embryos were stored in the petri dish in a hot water bath at 28.5°C inside a laminar flow hood, under natural lighting conditions until experimentation occurred at 3 dpf (**Pic. 2**).

Picture 2: This photo shows the laminar fume hood that the hot water baths were stored in. The embryos were housed in petri plates within the hot water baths set at 28.5°C.

TBI Inducing Methods
During pre-trials we determined that treatment in a shaking incubator at 250 rpm resulted in the optimal mortality between 10-25% with little to no morphological damage (**Pic. 3**). In humans, 10-25% of cases involving TBI results in mortality so we used this percentage as our baseline (Prins *et al.*, 2013). To induce TBI, we transferred an average of 14 embryos at 3 dpf into a cell culture tube containing 1 mL of

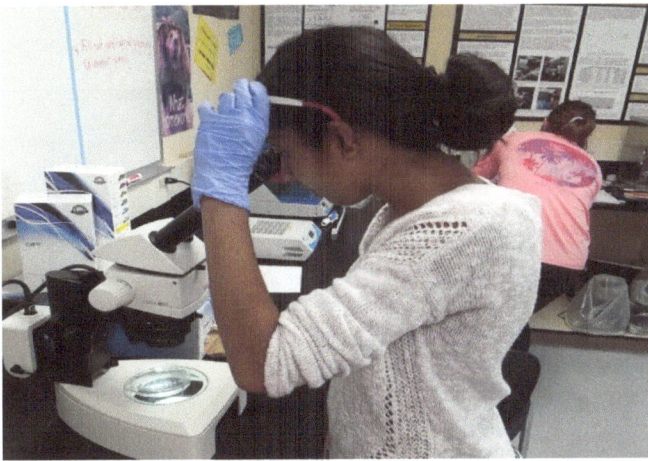

Picture 3: In this image Elango is using a Leica KL300 LED stereomicroscope to record the number of deceased embryos after treatment in the shaking incubator at 250 rpm.

egg water. All embryos had emerged from their chorion prior to treatment. The culture tube with the embryos was placed into the shaking incubator set to 250 rpm at 28.5 °C for 1 minute. Immediately following the treatment, the embryos were transferred into a six well plate to separate the control and experimental groups. Each well contained 5 mL of egg water and was returned to the hot water bath. All embryos in the no treatment control group received the same treatment except that they were not placed into the shaking incubator at 250 rpm. In order to measure the effects of the exposure to the TBI inducing event, we utilized two assays (response to touch and head size) following treatment to test for evidence of secondary symptoms of TBI.

Touch Assay

In previous literature, touch assays have been commonly used to assess muscle performance and neurological health by analyzing the reaction to a stimulus (Sztal, Ruparelia, Williams, & Bryson-Richardson, 2016). In this investigation, a touch assay was used as a measure of neurological damage with a lowered response to stimulus expected with head trauma. The touch assay was performed 1 day after TBI treatment in the shaking incubator for trial one and immediately after induction for trial two to account for potential differences to response to the touch with time. To perform the touch assay, we transferred the embryos from the six well plate into a 10 cm petri plate containing 1 mL egg water and used sharp tipped forceps to lightly tap the embryos on their head twice. We rated the reaction of the embryos to the stimulus on the second touch using a scale of one to three: one = little to no reaction to the touch, two = a clear reaction to the touch, and three = exaggerated and prolonged reaction to the touch. To avoid bias, the touch assays were rated blindly, where two team members performed and filmed the assay and the third researcher viewed the films, and assigned the rating without knowing which treatment the embryo received.

Head Size Assay

The second assay we used to measure TBI symptoms was a head size assay. This was performed to determine if measurable edema would occur with TBI treatment. One of the most significant and dangerous effects of a severe TBI is brain swelling. Head edema has been documented in adult zebrafish with TBI treatment (Solin, 2015). To measure edema, the head size of the embryos at 24 hours post TBI treatment (4 dpf) was recorded using the EVOS-FL Cell Imaging System Microscope (**Pic. 4**).

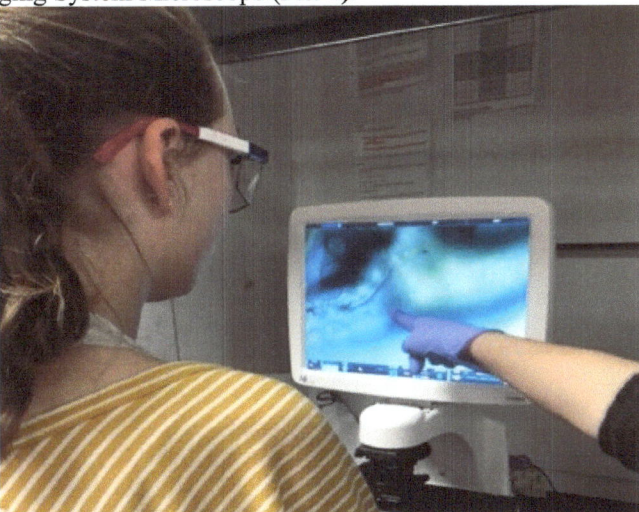

Picture 4: Kozlowski using the EVOS-FL microscope to image the zebrafish to prepare for the head size analysis.

To measure the head size, we anesthetized the embryos using tricaine administered to the egg water until they were immobilized. The embryos were imaged using the microscope with the LPlan 2x/0.06 objective under the bright field light setting. Embryos were placed in drops of glycerol laterally and measured the length of the forebrain, midbrain, hindbrain in micrometers using the scale bar (**Pic. 5**). The length of the entire body was also measured and used to normalize the data creating a head size/body size proportion for data analysis. This reduced variation and bias in the data due to embryo size.

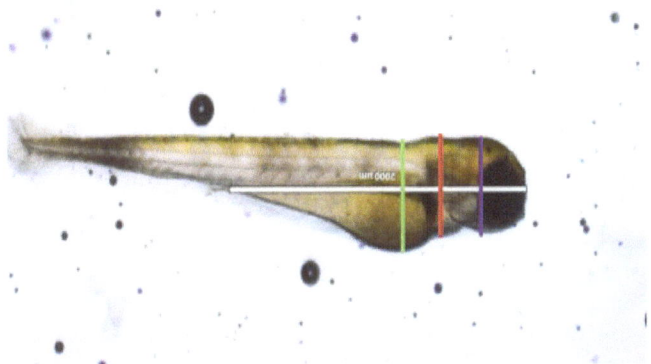

Picture 5: The head size measurements of the embryos were taken as followed: forebrain = purple, midbrain = red, hindbrain = green. The measurements of each individual section were proportioned with the entire body length.

Data Analysis

We analyzed the data using a paired t-test for the head size assay and a two proportion z-test for the touch assay. After performing all assays, we euthanized the embryos at 5 dpf following NIH guidelines which included freezing them at -20°C for one week and then disposing of them as biological waste (Matthews & Varga, 2012).

RESULTS

In this research, zebrafish embryos were tested to see if they could effectively serve as a model of the secondary injuries of TBI. The embryos were exposed to TBI inducing treatment in a shaking incubator at 250 rpm on 3 dpf and the subsequent response to stimulus in a touch assay as well as the head size was measured (**Pic. 6**).

Touch Assay

In the control and experimental group in both trials, the majority of embryos exhibited a rating of one, shown in **Graph 1**, which is indicative of very little to no response to the touch. The resulting data was analyzed with a two proportion z-test at a significance of 0.05. There was no statistically significant difference between the treatment and control group on the touch assay (trial 1 p value=0.9056, trial 2 p value=0.3265) which signifies that the TBI inducing treatment did not result in a difference in the touch assay.

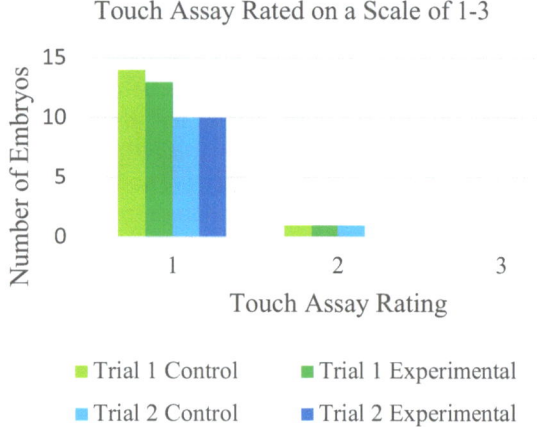

Graph 1: This graph depicts the number of embryos that were rated either a 1, 2, or 3 in the response to touch assay for both the treatment and control groups.

Head Size Analysis

In both trials, the differences in the forebrain, midbrain, and hind brain between the control group and experimental group was miniscule which demonstrated little head edema. The forebrain, midbrain, and hindbrain were measured and proportioned to the body length. **Graph 2** shows the average brain to body proportion between the control and experimental groups. The data was analyzed with a two proportion z-test at a significance of 0.05. The difference was not statistically significant as the resulting p-values ranged from 0.07 to 0.75, indicating that the shaking incubator did not affect head sizes.

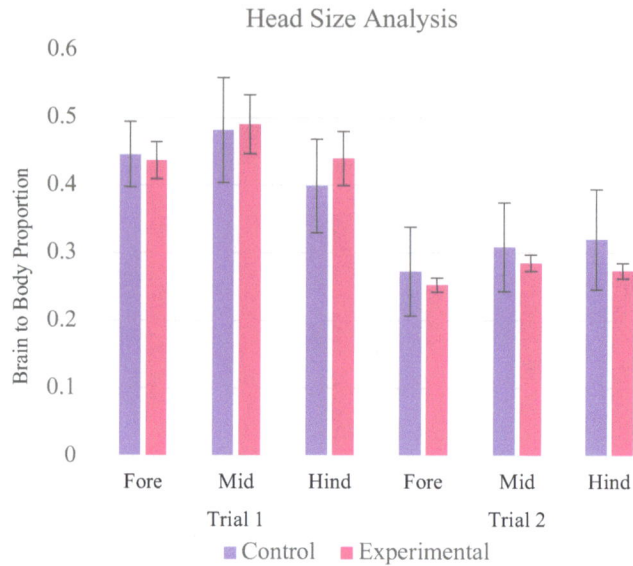

Graph 2: This graph depicts the average head size of zebrafish embryos post TBI inducing treatment as a proportion of head size to body length (μm). The fore, mid, and hind brain regions were measured for both the TBI inducing treatment and control groups.

DISCUSSION

As the number of TBIs continues to rise, the need for lower model organisms for early stage research and drug discovery is crucial. Without cheaper alternatives to higher model organisms, cost effective drug screens cannot be performed. Prior to this research, zebrafish embryos had not been fully investigated as a potential model organism of TBI, despite the fact that their use could contribute to future of potential treatments for TBIs. Our research investigated if zebrafish embryos could be used as a model organism for TBI research. We hypothesized that zebrafish embryos would serve as an effective model organism of the secondary injuries related to TBI including delayed neurocognitive response and brain edema. We utilized a touch and head size assay to determine if zebrafish embryos would exhibit a lowered response to stimulus or an increase in head size (evidence of edema) with TBI inducing treatment. To optimize the TBI inducing treatment, we first ensured that the mortality was between 10-25%. With this treatment, the touch assay failed to produce statistically significant changes in the embryos response to touch ($p = 0.55$) with both the treatment and control group averaging a 1 on the touch assay rating scale. In addition to this, the head size assay also failed to produce data that was different between the treatment and control groups. The differences between the control and experimental groups forebrain, midbrain, and hindbrain in trial 1 was 0.00876, 0.00913, and 0.04048 respectively. In trial 2 the differences between the control and experimental in the forebrain, midbrain, and hindbrain were 0.0198, 0.02346, and 0.04597 respectively. The p values for the head size assay ranged from 0.0747-0.749. Our results showed no significant difference between any of the control and treatment groups in both of the assays which means that treatment in a shaking incubator at 250 rpm is not an adequate method to induce secondary symptoms of TBI

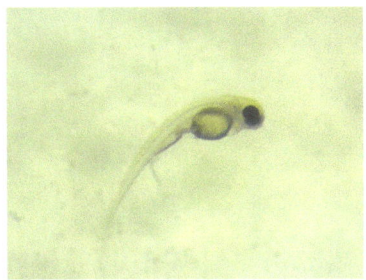

Picture 7: An example of damaged embryo caused by the shaking incubator.

delayed response to touch and head edema in zebrafish embryos.

As the embryos are very small and fragile, the large force created by the shaking incubator did not project a targeted force directly to the head of the embryo. Instead, it caused trauma to the entire body, which threatened to damage more than just the brain (**Pic. 7**). At higher rpms, which were tested prior to experimentation, it is possible that TBI symptoms could cause trauma and high mortality to the embryos. Any method to induce TBI that is not specific to only the brain is most likely going to be unsuccessful in zebrafish embryos due to the delicacy of their bodies. An embryo brain is only protected by an undeveloped larval cartilage structure. However, adult zebrafish have a developed cranial bone structure similar to the skull of a human, hence why adult zebrafish currently are used for TBI research (Mork & Crump, 2015). The zebrafish embryos cranial structure more closely resembles the fragility of a newborn baby. In future studies, zebrafish embryos may be used as a model for brain injuries found in newborns, such as fetal strokes that potentially lead to chronic diseases such as cerebral palsy (Traumatic Brain Injury.com, 2006). Complications occurring with childbirth can cause many lifelong neurological impairments.

A limitation of our research was the ability to determine if the zebrafish embryos used their genetic regenerative properties to respond to the injury. In our research, the embryos recovered for one day from the treatment before we performed the head size assay. It would be good to investigate if differences could be observed immediately after the trauma inducing event. It is possible that edema was produced with our method but the embryos were able to recover from the neurological harm in a short time frame. In future studies, analyzing the neuroregeneration in embryos would allow us to understand the cellular processes that occur in response to injury. The use of zebrafish adults in neuroregenerative studies shows the capability of embryos as a similar model, and translating that model to embryos would allow for more research centered on the problems with infant brain injury. Nonetheless, zebrafish embryos do show potential to model TBI for the regenerative properties but the method to induce these symptoms still needs to be elucidated.

In addition to not being able to identify if the embryos were using their regenerative properties, the touch and head size assays relied on the secondary injuries of TBIs which varied between each case. The use of zebrafish embryos for this research would require understanding of the chemicals of the brain. A more accurate way to determine neuronal damage from a TBI would require identifying inflamed cytokines in the brain. With limited laboratory equipment, we were unable to explore direct neuronal damage or measure cytokine inflammation within the extracellular space of the

zebrafish brain (Thelin *et al.*, 2018). Using biomarkers to detect the damage could have affected the outcome of our research and better determine if the shaking incubator caused damage to the brain. In humans, besides a CAT or MRI scan, physical symptoms are solely used as a sign for brain injury due to a lack of technology to detect brain damage. Physical damage to the brain is visible with these scans but still lacks the ability to see neuronal damage. The assays we performed mimicked similar symptom tests doctors use on TBI patients as currently they are some of the only few physical tests done to detect brain damage.

The main source of error in this experiment was our lack of experience with breeding zebrafish, which limited us in the number of embryos available for our trials. In addition to errors in the breeding process, determining the best method to anesthetize and position the embryos for imaging was challenging and may have led to inconsistencies in our measurement of the head size.

ACKNOWLEDGMENTS
We would like to thank Katie Zilligen for being a part of this research by assisting us in taking care of the fish and as well as contributing her knowledge and time throughout the experiment. We would like to thank Karlie Fedder and Kayt Scott for mentoring us throughout our project and helping us troubleshoot problems we encountered. We would like to thank our Biotechnology teacher Shawndra Fordham for mentoring us in all aspects of this research. We greatly thank the Kozlowski, Elango, DeMarte, and Ashbeck families for financially funding this project, as well as Nayan Naik and the Rock Canyon High School science department. We would like to thank the following RCHS teachers: Gwendolyn Karaba for aiding us in determining how to collect and analyze our data for statistical analysis, Kerry Hinton for ensuring that the chemicals that we used in this project were safe to work with, Bryan Winkelman for assisting us with setting up our website and assisting us in publishing this journal article. We would also like to thank Kristin Karnicki and Peter Thompson for giving us insight and feedback on how to improve our project. We would like to thank Rock Canyon High School and Douglas County School District for granting us the lab space, equipment, and partial funding for this research. We would not have been able to complete this research had it not been for them. Lastly, we would like to thank Audrey Gruszczynski, Lauren McCaffrey and Molly Dolan, former RCHS Biotechnology Research students, for helping us edit our paper.

REFERENCES
Bullock, G. (2018). Long Term Effects of Post-Concussion Syndrome. Retrieved from https://www.theraspecs.com/blog/long-term-effects-post- concussion- syndrome/

Cacialli, P., Palladino, A., & Lucini, C. (2018). Role of brain-derived neurotrophic factor during the regenerative response after traumatic brain injury on adult zebrafish. *Neural Regeneration Research, 13*(6), 941-944. doi:10.4103/1673-5374.233430

Chitramuthu, B. (2013). Modeling Human Disease and Development in Zebrafish. *Human Genetics & Embryology, 3(*1). doi:10.4172/2161-0436. 1000e108

Johnson, V., Stewart, W., & Smith, D. (2013). Axonal Pathology in Traumatic Brain Injury. *Experimental Neurology, 246*, 35-43. doi:10.1016/j. expneurol.2012.01.013

Katz, D., Cohen, S., & Alexander, M. (2015). Mild Traumatic Brain Injury. *Handbook of Clinical Neurology, 127*(9), 131-156. doi.org/10.1016 /B978-0-444-52892 -6.00009-X

Kimmel, C., Ballard, W., Kimmel, S., Ullmann, B., & Schilling, T. (1995). Stages of Embryonic Development of the Zebrafish. *Developmental Dynamics, 203*(3). 253-310. doi: 10.1002/aja.1002030302

Lynch, P. (2006). Skull and brain normal human [Digital Image] Retrieved from://commons.wikimedia.org/wiki/File:Skull_and_brain_normal_human.svg.

Maheras, A., Dix, B., Carmo, O., Young, A., Gill, V., Sun, J., ... Spence, R. (2018). Genetic Pathways of Neuroregeneration in a Novel Mild Traumatic Brain Injury Model in Adult Zebrafish. *eNeuro, 5*(1). doi:10.1523/eneuro.0208-17.2017

Matthews, M. & Varga, Z. (2012). Anesthesia and Euthanasia in Zebrafish. *Journal of the Institute for Laboratory Animal Research, 53*(2).192-204. doi:10.1093/ ilar.53.2.192

Mork, L. & Crump, G. (2015) Zebrafish Craniofacial Development: A Window Into Early Patterning. *Current Topics in Developmental Biology, 115.*235-260. doi:10.1016/bs.ctdb.2015.07.001

Murthy, T., Bhatia, P., Sandhu, J., Prabhakar, T., & Gogna, R. (2005). Secondary Brain Injury: Prevention and Intensive Care Management. *Indian Journal of Neurotrauma, 2*(1), 7-12. doi:10.1016/S0973-0508(05)80004-8

National Institute of Health. (2016) Guidelines for Use of Zebrafish in the NIH Intramural Research Program, NIH, Intramural Research Program. (2016). [PDF file] Retrieved from https://oacu.oir.nih.gov/sites/default/files/uploads/arac-guidelines/zebrafish.pdf

Prins, M., Greco, T., Alexander, D., & Giza, C. (2013). The pathophysiology of traumatic brain injury at a glance. *Disease Models & Mechanisms,6*(6). 1307-1315. doi: 10.1242/dmm.011585

Protect the Brain. (2018). What is a Concussion? Retrieved from http://www.protectthebrain.org/Brain-Injury-Research/What-is-a-Concussion-.aspx

Rudy, S., Maia, P., & Kutz, J. (2016). Cognitive and behavioral deficits arising from neurodegeneration and traumatic brain injury: a model for the underlying role of focal axonal swellings in neuronal networks with plasticity. *Journal of Systems and Integrative Neuroscience, 2*(1), 114-121. doi:10.15761/JSIN.1000120

Schmidt, R., Beil, T., Strähle, U., & Rastegar, S. (2014). Stab wound injury of the zebrafish adult telencephalon: a method to investigate vertebrate brain neurogenesis and regeneration. *Journal of Visualized Experiments : JoVE, 1*(90), e51753. doi:10.3791/51753

Solin, S. (2015). *Modeling pediatric brain and central nervous system cancer in zebrafish* (Doctoral dissertation). Retrieved from https://lib.dr.iastate.edu/etd/14910

Spinal Cord (2018). Types of Traumatic Brain Injury. Retrieved from https://www.spinalcord.com/types-of-traumatic-brain-injury

Sztal, T., Ruparelia, A., Williams, C., & Bryson-Richardson, R. (2016). Using Touch-evoked Response and Locomotion Assays to Assess Muscle Performance and Function in Zebrafish. *Journal of Visualized Experiments,* (116). doi:10.3791/54431

Taylor, C., Bell, J., Breiding, M., & Xu, L. (2017). Traumatic Brain Injury-Related Emergency Department Visits, Hospitalizations, and Deaths -United States, 2007 and 2013. *Surveillance Summaries, 66*(9). 1-16. doi: 10.15585/mmwr.ss6609a1

Thelin, E., Hall, C., Gupta, K., Carpenter, K., Chandran, S., Hutchinson, P., ... Helmy, A. (2018) Elucidating Pro-Inflammatory Cytokine Responses after Traumatic Brain Injury in a Human Stem Cell Model. *Journal of Neurotrauma 35.* 341-352. doi:10.1089/neu.2017.5155

The Zebrafish Information Network. (2013). General Methods for Zebrafish Care. Retrieved from https://zfin.org/zf_info/zfbook/chapt1/1.3.html

Traumatic Brain Injury.com. (2006). Birth Trauma. Retrieved from http://www.traumaticbraininjury.com/birth-trauma/

Unterberg, A., Stover, J., Kress, B., & Kiening, K. (2004). Edema and brain trauma. *Neuroscience 129*(4), 1019-1027. doi:10.1016/j.neuroscience. 2004.06.046

Your Genome. (2014).Why use the zebrafish in research? Retrieved from https://www.yourgenome.org/facts/why-use-the-zebrafish-in-research

Zebrafish Health. (2010). Zebrafish Facts. Retrieved from http://www.zf-health.org/information/factsheet.html

ABOUT THE AUTHORS

Pictured: Pictured from left to right are Kozlowski, Elango, and their mentors from CU Anschutz Medical Campus, Karlie Fedder and Kayt Scott.

Throughout this research, Kozlowski and Elango have experienced many setbacks and life lessons that they each learned from. They learned perseverance and patience as many things did not turn out the way that they had expected. They learned how to prepare for mistakes and how to manage their time. Anything science related always takes much longer than planned, and they both were able to experience that first hand. They developed their teamwork skills and learned to work with many different people in many different situations. The time that Kozlowski and Elango spent in the lab allowed them to develop many skills, especially working with zebrafish embryos. This research experience will open up many doors to their future careers.

Kozlowski loves anything math or science, which prompted her to take the challenge that is the Rock Canyon High School Biotechnology Program. She took Introduction to Biotechnology which introduced her to the Experimental Design in Biotechnology course. She has taken Honors Biology, Honors Chemistry, and is planning on taking AP Physics I & II. In addition, she is involved with Science National Honor Society. She has a passion for neuroscience and biomedical sciences and she hopes to incorporate both into her future education. The experiences and lessons she has learned in this class have improved not only her science skills, but also have contributed a great deal to her personal growth. She was exposed to a variety of different difficult situations that helped her expand her patience, communication, and problem solving.

Elango is very passionate about anything science related. She has taken many upper level science classes including AP Chemistry, Foundations of Physics, Honors Biology, Honors Chemistry, and is planning on taking AP Physics C next year to expand her knowledge of science. Prior to Experimental Design in Biotechnology, she was in the Introduction to Biotechnology course and decided to explore the field of research more. She has been a part of Science Olympiad for the past three years, Chemistry Olympiad for the last year, and is also a part of Science National Honor Society.. Throughout this course, Elango has been able to experience the realities of what it is like to design her own research project and how to deal with the struggles as well. Along the way, she has learned many valuable life lessons in working with different people, advocating for herself, and communicating, which have impacted her tremendously.

Inducing neuronal injury in *Caenorhabditis elegans* to serve as a model organism in traumatic brain injury research

Z. Zizzo, S. F. Bermingham, T. J. Tankersley, & S. M. Petri
Department of Science, Principles of Experimental Design in Biotechnology, Rock Canyon High School, Highlands Ranch, Colorado, USA

Traumatic Brain Injury (TBI), indicated by neuronal damage, is caused by acute head trauma. Despite the lasting damage of a TBI, there is currently a lack of medication to treat neuronal damage directly. To better treat TBI, increased understanding of the mechanisms of neuronal damage and treatments that promote healing and repair is needed. In our research, protocols were engineered and verified for inducing neuronal damage in a transgenic strain of [F49H12.4::GFP + unc-119(+)] *Caenorhabditis elegans* with GFP labeled neurons. We hypothesized that it is possible to induce measurable and statistically significant damage. The treatment group received treatment in the shaking incubator at 300 rpm for 5 minutes in a microcentrifuge tube with two beads. To measure neuronal damage, neurons in the *C. elegans* were visualized using an EVOS FL microscope and rated using a 1-5 Neuronal Damage Severity Scale. The average ratings for the control and treatment groups were 1.55 and 3.95 respectively, with a lower number representing less damage. The treatment group, when analyzed using a t-test, demonstrated a statistically significant difference from the control with a p-value of 0.0000134163. This means *C. elegans* treated with a shaking incubator can be an effective model of TBI neuronal damage. This research is a remarkable first step to the development of treatments for the neuronal damage associated with TBI. In the future, protocol optimization in order to consistently induce a range of severities is necessary for use in future drug studies.

Brain injuries are a dangerous and growing issue across the globe. In the United States alone, there is an occurrence of a TBI every 15 seconds, resulting in 1.7 million new head injury cases (Prins, Greco, Alexander, & Giza, 2013). Approximately 50,000 deaths in America each year are due to TBI. TBI is most commonly seen in athletes under the age of 19 and military personnel. TBI occurs after a jolt or blow to the head which disrupts the neuronal makeup of a person's brain. Repeated brain trauma is connected with life-altering and deadly diseases, including Parkinson's, Alzheimer-like dementia, motor neuron disease, and chronic traumatic encephalopathy. Additionally, the development of mental disorders, lower memory index and cognitive function, frequent headaches, and loss of hearing, are all long-term effects that can occur with repeated brain injury (Oyegbile, Delasobera, & Zecavati, 2018). Incidents of TBI are seen in people of all ages but are becoming increasingly prevalent in youth, especially student-athletes. While not every TBI leads to permanent brain damage, potential short-term side effects include headaches, unsteadiness, difficulty learning, mood changes, confusion, disorientation, slowed thinking, nausea, and loss of consciousness lasting up to 30 minutes. Along with this, TBI may cause amnesia up to 24 hours after trauma is induced as well as memory loss; both short and long-term. Each TBI is unique and memory loss can be temporary for some, but long-term for others (Dennis, 2009).

Concussions, a form of mild TBI, are a global concern as approximately five million people suffer from them each year (Head Case Company, 2013). Similar to TBI in general, the rate of concussions is rising each year with 1.1 to 1.9 million sport-related youth head injuries annually (Bryan,

Rowhani-Rahbar, Comstock, & Rivara, 2016). In the 2015-2016 school year, a study conducted by the Michigan High School Student Athlete Association found 4,452 sport-related concussions were diagnosed in participating schools, which was the equivalent to six injuries per high school during that one year period (Kimmerly, 2016). A similar study conducted in Canada found approximately 20% of teens in grades 8, 10, and 12 received at least one concussion diagnosis (Veliz, McCabe, Eckner, & Schulenberg, 2017). A person's first concussion usually does not cause permanent damage, depending on the severity, but subsequent head

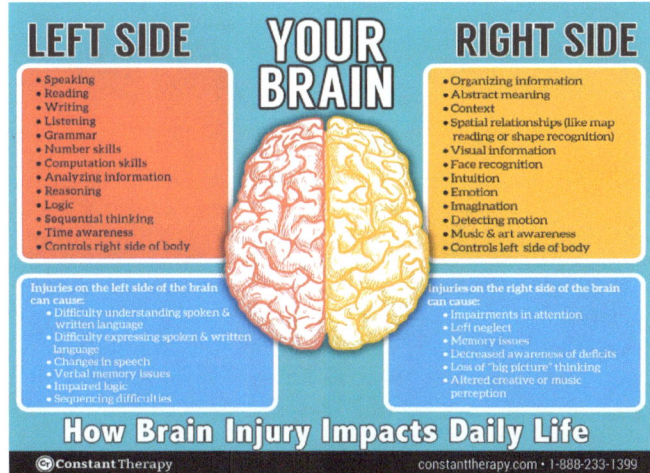

Figure 1: This graphic depicts the impacts a TBI can have on the left and right side of the brain. From "Right Brain Injury vs. Left Brain Injury Understanding the Impact of Brain Injury on Daily Life" by Constant Therapy, The Learning Corp (https://thelearningcorp.com/). Reprinted with permission.

injuries can result in lasting effects ranging from permanent disability to death (American Association of Neurological Surgeons, 2018) (**Fig. 1**). A TBI even as mild as a concussion can truly alter a season of someone's life or the course of their life as a whole. From something as simple as a blow to the head from a basketball to a fractured skull after a diving accident, these head injuries result in great amounts of stress for a person and their family. Falling behind in school or work and experiencing difficulty with learning or memory are common after a concussion. For instance, two of our three research team members have had to endure the immense hardship of missing school, loss of memory, and overall confusion due to their personal experience with having a concussion.

On a cellular level, when a TBI event occurs, the nerve cells in the brain are directly damaged. After a traumatic impact, blood flow to the brain is either increased (hyperfusion) or decreased (hypoperfusion). When this occurs there is often an impairment to the cerebral metabolic function, cerebral oxygenation, and cerebrovascular autoregulation. The damage nerve cells undergo due to these complications leads to inflammation and ultimately cell death (Werner & Engelhard, 2007). Additionally, when a TBI occurs, axonal shearing can cause neurons to misfire. In this type of damage, the axon disconnects from the cell body which can prevent cells from delivering signals, rendering them useless (**Fig. 2**). TBI can cause demyelination of neurons, also affecting their ability to effectively transmit signals (Centre for Neuro Skills, 2019). With more research, an increased understanding of the neuronal damage associated with TBI would allow for the development of medications to specifically treat the direct neuronal damage that occurs.

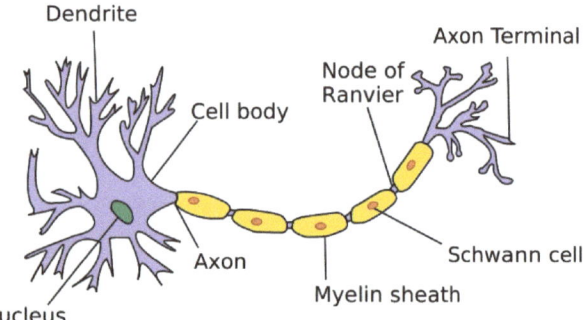

Figure 2: The above figure is a labeled diagram of the human neuron. Reprinted from "Anatomy and Physiology" by the US National Cancer Institute's Surveillance, Epidemiology and End Results (SEER) Program. CC By SA 3.0

In order to treat TBI, there is a growing need for a better understanding of the cellular mechanisms involved. Invertebrate model organisms are necessary for performing early-stage research to advance this understanding and develop treatments for this devastating condition. In this research, our goal was to engineer a method to induce visible and quantifiable neuronal damage in *C. elegans*. We hypothesized that we would be able to induce neuronal damage in *C. elegans* using a shaking incubator and 1.00 mm glass beads. We used the transgenic NC1686 strain of *C. elegans* with GFP tagged neurons that are observable under a fluorescent microscope. Successful induction of neuronal damage was visualized with swollen or broken neurons. Neuronal damage was rated between 1 to 5 on a Neuronal Damage Severity Scale in a blind test, with a 1 indicating no swelling/neuronal breaks and a 5 indicating severe swelling and neuronal breaks.

C. elegans were chosen for this research because they have the potential to be an excellent model organism associated with TBI as induced neuronal damage can be visualized with trauma. *C. elegans* are microscopic, transparent, non-parasitic, hermaphroditic nematodes that are great for use in research laboratories as they are low maintenance, inexpensive, and easy to handle and maintain. Stored in the laboratory at 18-22°C, *C. elegans* are housed in petri plates on NGM agar and fed OP50 *Escherichia coli*. They produce numerous offspring, up to 300, in a very short four day life-cycle with a two to three week life-span (**Fig. 3**). Their transparent body simplifies studying and visualizing cellular structures and organs. The *C. elegans* most complex organ is its neuronal system, which is well mapped and categorized. While *C. elegans* do not have brains, they have a well-developed neuronal network of ganglia surrounding the pharynx. When visualized under a microscope, all 302 neurons are visible (Edgley, 2015). Along with easy visualization, *C. elegans'* ability to absorb small molecular drugs topically or orally also make them ideal for drug research studies.

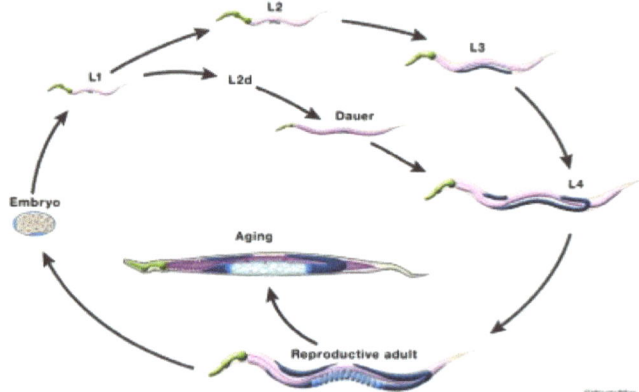

Figure 3: This figure depicts the life cycle of *C. elegans,* a microscopic nematode. From Herndon, L.A., Wolkow, C.A., Driscoll, M. and Hall, D.H. 2018. Introduction to Aging in *C. elegans*. Reprinted from Wormatlas (wormatlas.org) with permission.

METHODS

This experiment assessed the ability of *C. elegans* to serve as a model organism for future TBI research as a direct measure of visible neuronal damage. In this protocol we used the NC1686 transgenic strain of *C. elegans* with GFP labeled neurons [F49H12.4::GFP + unc-119(+)]. Following the treatment to induce neuronal damage, the neurons of the *C. elegans* were visualized using an EVOS FL microscope with the GFP cube at 40X magnification.

Care and Maintenance

Throughout the experiment, *C. elegans* were housed on NGM-Lite petri plates seeded with OP50 *E. coli* and stored in an incufridge at 22°C. Prior to performing experiments on the *C. elegans*, we synchronized the age of the population by picking gravid adults directly into a drop of sodium hypochlorite which consisted of 0.67g NaOH, 6.67 mL of household bleach, and 25 mL distilled water. A second drop was placed on top to ensure sufficient exposure to the bleach solution (Hupka, Rohr, Samuel, & Fordham, 2018). This synchronization protocol dissolved adult *C. elegans* but left the embryo intact because of the protective membrane.

Inducing Neuronal Damage

Four day old adult *C. elegans* were placed into microcentrifuge tubes containing 5 µL of 1% 1-phenoxy-2-propanol anesthetic in M9 buffer along with two 1 mm glass beads. The treatment group received TBI inducing treatment at 300 rpm in a shaking incubator at 22°C for 5 minutes after which they were immediately visualized under the EVOS FL microscope (**Pic. 1**). The control group was also placed into microcentrifuge tubes containing the anesthetic, however, they were not exposed to the shaking incubator treatment and instead remained in the incufridge at 22˚C.

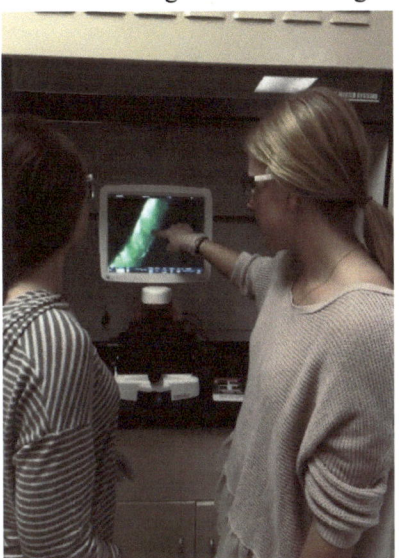

Picture 1: In this image, Bermingham and Tankersley visualize the neuronal damage induced with the shaking incubator treatment using the EVOS microscope analyzing the *C. elegans'* neuron flouresced with the GFP protein.

Visualizing Neuronal Damage

Damage can be visualized in *C. elegans* in either the form of breaks or swellings. Breaks are more sev-ere forms of neuronal damage whereas swel-lings are less severe. Directly after inducing neuronal damage, the *C. elegans* were imaged on the EVOS FL inverted microscope using the GFP cube and a long distance working EVOS Inf Plan Fluor 40X, 0.75NA/0.72WD objective. Agar pads at a 10 ug/mL concentration were used for visualization. Two microscope slides were covered in one layer of masking tape to regulate the thickness of the agar pad and placed on either side of the slide containing the agar pad where an additional glass slide was placed on top of the agar to flatten the pad (**Pic. 2**). Approximately 30 *C. elegans* from each treatment and control group were mounted in the agar pad and the central neuron was visualized in the medial region.

Picture 2: In this picture, Zizzo makes the agar pad used to mount *C. elegans* for visualization on the EVOS FL microscope.

Neuronal Damage Scale

To quantify the damage, images of the neuronal network in the medial region of the *C. elegans* were analyzed using the Neuronal Damage Severity Scale (**Fig. 4**). Each *C. elegan*s was rated between a 1 to 5 where a rating of 1 was assigned when no visible neuronal breaks or swellings were present, a clear central neuron. A rating of 2 was assigned when minor swellings but no breaks were evident. A rating of 3 was

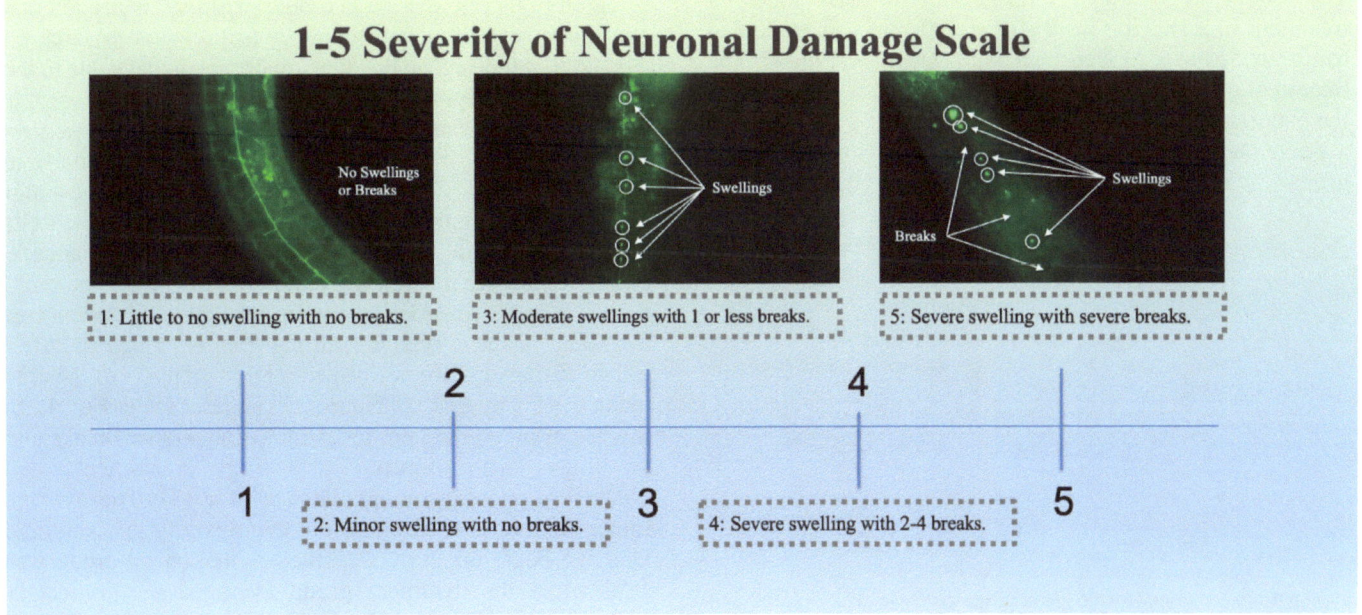

Figure 4: The above severity scale was used in an unbiased, blind rating to examine *C. elegans* neuronal health in both the control and treatment groups. The three reference pictures illustrate what we consider swellings (less severe damage) and breaks (more severe damage) in a *C. elegans'* neuron.

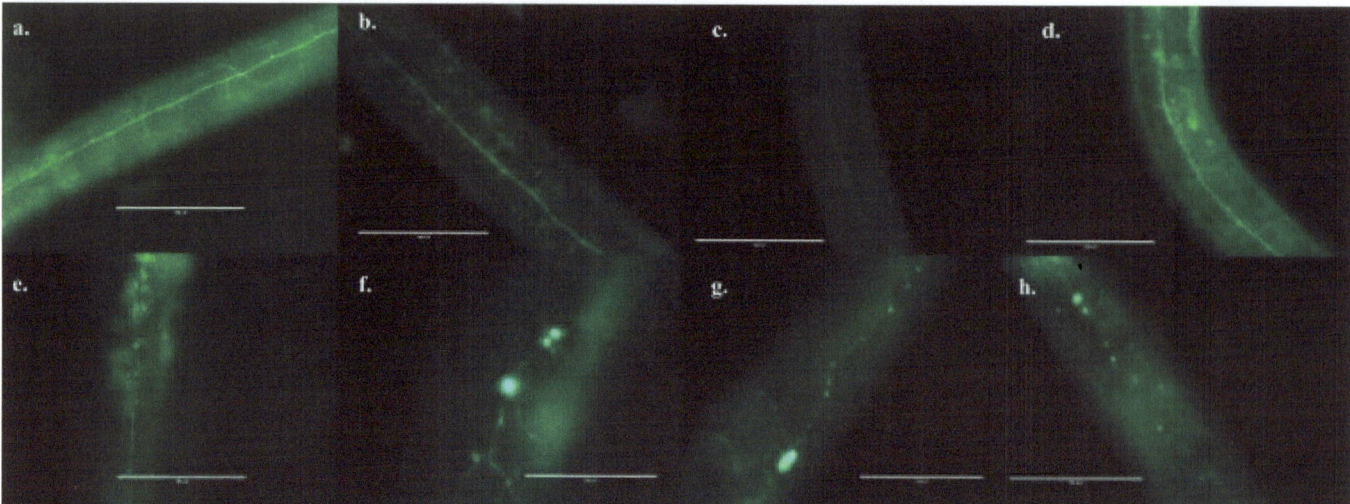

Picture 3: These images display the range of damage found in the no TBI treatment control and the TBI treatment groups. Images **a-d** are from the control group and all were rated as a 1 on the Severity of Neuronal Damage Scale. Images **e-h** are from the TBI treated group with e and f rated as a 4 and **g-h** rated as a 5 on the scale.

assigned with moderate swellings and one or fewer breaks, while a rating of 4 was used when severe swelling and four to five breaks were visible. Last, a rating of 5 was used when the central neuron displayed extreme amounts of swelling with severe breaks to the point where the central neuron was no longer intact. Images where the neurons were not clear enough to accurately rate the damage were not used in our experiment. A total of 21 images were selected and rated on the scale at random from both the treatment and control groups. In an effort to reduce bias, ratings were performed blindly where the two researchers assigning the ratings did not know whether the image was from the treatment or control group. Data from all images were analyzed using a paired t-test to determine the statistical significance of the results.

RESULTS

The extent of neuronal damage due to TBI inducing treatment in a shaking incubator at 300 rpm was evaluated using the Severity of Neuronal Damage Scale (**Fig. 4**). The treatment group typically scored a 4 or a 5 on the scale while the majority of the control group scored a 1 or 2 (**Pic. 3**). The average neuronal damage rating for the no treatment control group was 1.55 with the majority of the *C. elegans* exhibiting

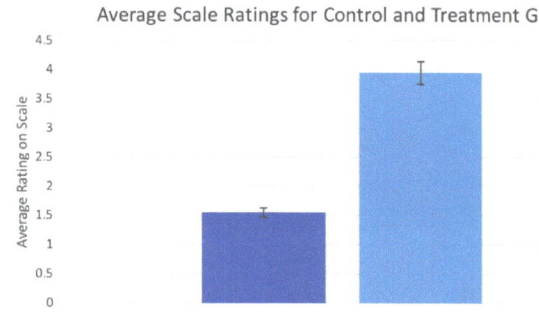

Graph 1: Pictured above is a graph of the averages of our control and experimental group data with five percent error bars. Data was collected using an unbiased scale ranking.

minimal evidence of neuronal swelling and breaks. In contrast, the average rating of the *C. elegans* that had been exposed to treatment was 3.95 (**Graph 1**). When compared using a t-test, the resulting p-value of 1.34163×10^{-5} demonstrates that there was a statistically significant difference between the treatment and control groups in the extent of neuronal damage in the *C. elegans* on the Severity of Neuronal Damage Scale.

DISCUSSION

TBI is an extremely prevalent problem worldwide, yet the medications available to directly target and treat the neuronal damage present following TBI are limited. This lack of treatment options could be improved with an increase in research and understanding of the cellular and molecular mechanisms involved with direct damage to the neurons that occurs with a TBI. To perform early-stage discovery-based research in this field, a good model organism is needed that can effectively simulate this injury as previous research on TBI has not focused on the direct neuronal damage due to the difficulty quantifying it.

This research was conducted to determine if *C. elegans* could serve as a model organism of the neuronal damage associated with TBI for future research. A shaking incubator was tested as a method to induce neuronal damage due to its ease of use and accessibility in a laboratory environment. Glass beads were incorporated into the centrifuge tubes with the *C. elegans* in order to induce an increased amount of neuronal damage without killing the *C. elegans*. After inducing damage, we performed a blind analysis of the data using the 1-5 Severity of Neuronal Damage Scale (**Fig. 4**). In this experiment, the neurons demonstrated clear breaks and swellings with injury, al-lowing assessment and quantification of neuronal damage (**Pic. 3**). The average rating of neuronal damage on the Severity of Neuronal Damage Scale for the *C. elegans* in the control group was 1.55 while the treatment group averaged a significantly higher 3.95 (**Graph 1**). Not only were we able to demonstrate that with treatment in a shaking incubator, visible and quantifiable neuronal damage can be produced, but also that this measurable damage was statistically significant. The resulting p-value of the paired t-test calculated on these averages was 1.34163×10^{-5}. Due to this exceedingly low p-value, we are able to conclude that *C. elegans* are a viable model organism for observing direct neuronal damage due to trauma and that treatment in a shaking incubator is a method for effectively inducing this damage. Using *C. elegans* as a model organism for TBI research is a viable option as using *C. elegans* allows scientists to directly visualize the neuronal damage post trauma.

Similarly, multiple research studies have found other model organisms to be effective models of human TBI. A previous study successfully used adult *Danio rerio* (zebrafish) as a TBI model by inserting a 27-gauge needle into the brain of the *D. rerio*. The *D. rerio* were examined over a 35 day time period, and a significant amount of cell proliferation was exhibited. The telencephalon region of *D. rerio* was found to be able to regenerate despite injury (Kishimoto, Shimizu, & Sawamoto, 2012). *Drosophila melanogaster* (fruit flies) have also been tested in TBI related research. With the use of an Omni Bead Ruptor-24 Homogenizer to induce head trauma, scientists analyzed the behavior of the fruit flies post damage inducing treatment (Barekat *et al.*, 2016). While TBI was successfully induced in both cases, it is difficult to directly visualize damage to the neurons in these organisms. When using *C. elegans*, however, it is possible to see the damage induced and track recovery of the neuron specifically due to the fact that *C. elegans* are transparent and have a sophisticated and well understood neuronal network.

The largest source of error in this investigation was initially caused by inexperience transferring the *C. elegans* to the microcentrifuge tubes, especially in the beginning stages of the research. The damage that was induced just by picking the *C. elegans* resulted in the outliers with higher ratings in our control group. As we proceeded with our research, we became more proficient at transferring the *C. elegans* without inducing damage which eliminated the outliers in later trials. With more data points, outliers could have been eliminated; however, due to time constraints, this was not possible. It is interesting to note, however, that simply picking *C. elegans* roughly can effectively induce neuronal damage. This would be a possible future study to determine if consistent and statistically significant damage could be produced through picking alone.

Despite the fact that this method was determined to produce a statistically significant amount of neuronal damage, modifications are necessary to optimize the protocol before using this method for research and early-stage drug screening. Data must be collected on a larger scale in order to eliminate outliers and to get more consistent results. Future research also needs to focus on the most consistent way of inducing and quantifying neuronal damage. Various shaking intensities or different time intervals should be experimented with to induce different ranges of severity in the *C. elegans* consistently, as it will better model the wide range of severities of a human TBI. In addition, an imperative future step would require other assays to be developed and tested in *C. elegans* that would enable researchers to relate quantified neuronal damage with other TBI symptoms, including learning and memory through touch, learning, or chemotaxis assays. TBI causes damage to a person's memory, response time, and learning ability; therefore, the use of assays is a significant future step as it will allow for scientists to investigate the relationship between neuronal damage and the symptoms associated with a TBI. Through this research, *C. elegans* have been shown to be an effective model of TBI using our tested trauma inducing method of treatment in a shaking incubator with the presence of glass beads. Once this protocol is further optimized, basic research into the cellular mechanisms as well as pharmaceutical screens can be performed.

ACKNOWLEDGMENTS
We would like to thank Dr. Christopher Link of CU Boulder, who mentored us and supported our research by providing us with training on using an EVOS microscope and mounting *C. elegans* for visualization, as well as important materials including the NC1686 transgenic strain [F49H12.4::GFP + unc-119(+)] of *C. elegans* with GFP labeled neurons, 1.00 mm and 3.00 mm glass beads, 1% 1-phenoxy-2-propanol anesthetic in m9 buffer, and 10 mm NGM lite pre-poured petri plates. Our research would not have been possible without the support of Biotechnology teacher Shawndra Fordham, who donated her time, instruction, and lab space for our work. We would also like to recognize Bryan Winkelman, RCHS Teacher Librarian, for reviewing our website blog posts, designing and editing our website, and for assisting with our poster and presentation. Additionally, thank you to former Biotechnology Research Students, Megan Hupka and Tabitha Samuel, who taught us many of the protocols we needed to learn to perform this research including seeding plates, making a synchronized population, and picking *C. elegans*. A huge thank you to Rock Canyon High School's Science Department and Nayan Naik for their generous grant to fund our project and to the Caenorhabditis Genetics Center for their generous donation of OP50 *E. coli* used throughout our research to feed the *C. elegans*. We also want to thank David Ferguson and Kerry Hinton, RCHS Chemical Safety Managers and Chemistry Instructors, for reviewing the safety of the chemicals we used in this project and ensuring our protocols were safe and Gwendolyn Karaba, RCHS Statistics Instructor, who supported us in our statistical analysis and quantification of our data. Lastly, we would like to thank Rock Canyon High School and the Douglas County School District for their ongoing support of our research and the RCHS Biotechnology Program.

REFERENCES

American Association of Neurological Surgeons. (2018). Concussion. Retrieved from http://www.aans.org/Patients/Neurosurgical-Conditions-and- Treatments/Concussion

Barekat, A., Gonzalez, A., Mauntz, R. E., Kotzebue, R. W., Molina, B., El-Mecharrafie, N., . . . Ratliff, E. P. (2016). Using Drosophila as an integrated model to study mild repetitive traumatic brain injury. *Using Drosophila as an Integrated Model to Study Mild Repetitive Traumatic Brain Injury*. doi:10.1038/srep25252

Bryan, M. A., Rowhani-Rahbar, A., Comstock, R., & Rivara, F. (2016). Sports- and Recreation-Related Concussions in US Youth. *Seattle Sports Concussion Research Collaborative, 138*(1). doi:10.1542/peds.2015-4635

Centre for Neuro Skills. (2019). Neuronal Firing. Retrieved from https://www.neuroskills.com/brain-injury/neuronal-firing.php

Dennis, K.C. (2009, August). Current Perspectives on Traumatic Brain Injury. Retrieved from https://www.asha.org/Articles/Current-Perspectives-on-Traumatic-Brain-Injury/

Edgley M. (2015). What is *C. elegans?* Retrieved from https://cbs.umn.edu/ cgc/what-c-elegans

Head Case Company. (2013). Sports Concussion Statistics. Retrieved from http://www.headcasecompany.com/concussion_info/stats_on_concussions_sports

Herndon, L.A., Wolkow, C.A., Driscoll, M. & Hall, D.H. (2018). Introduction to Aging in C. elegans. In WormAtlas. Retrieved from http://www.wormatlas.org/aging/introduction/mainframe.htm

Hupka, M., Rohr, S., Samuel, T., & Fordham, S. (2018). The effect of Piracetam on beta-amyloid plaque formation and subsequent paralysis on an Alzheimer's model *Caenorhabditis elegans*. *Rock Canyon High School Biotechnology, 3*, 47

Hutter, H. (2008). C. elegans nervous system. Retrieved from http://www.sfu.ca/biology/faculty/hutter/hutterlab/research/Ce_nervous_system.html

Kimmerly, G. (2016, September). MHSAA Announces 2015-16 Concussion Data from Yearlong Study. Retrieved from https://www.nfhs.org/articles/mhsaa-announces-2015-16-concussion-data-from-yearlong-study/#

Kishimoto, N., Shimizu, K., & Sawamoto, K. (2012). Neuronal regeneration in a zebrafish model of adult brain injury. *Disease Models & Mechanisms,* 200-209. doi:10.1242/dmm.007336

Oyegbile, T. O., Delasobera, B. E., & Zecavati, N. (2018). Postconcussive Symptoms After Single and Repeated Concussions in 10- to 20-Year-Olds: A Cross-Sectional Study. *Journal of Child Neurology, 33*(6), 383 - 388. https://doi.org/10.1177/0883073818759436

Prins, M., Greco, T., Alexander, D., & Giza, C. C. (2013). The pathophysiology of traumatic brain injury at a glance. *Disease Models and Mechanisms*, 6(6), 1307–1315. doi: 10.1242/dmm.011585

The Learning Corp. (2019) Constant Therapy. Right Brain Injury vs. Left Brain Injury | Understanding the Impact of Brain Injury on Daily Life. Retrieved from https://cdn2.hubspot.net/hubfs/3737327/Blog/brainrightleft.jpg

US National Cancer Institute's Surveillance, Epidemiology and End Results (SEER) Program. (2019). Neuron description. Retrieved from https://commons.wikimedia.org/wiki/File:Neuron.svg.

Veliz, P., McCabe, S. E., Eckner, J. T., & Schulenberg, J. E. (2017). Prevalence of Concussion Among US Adolescents and Correlated Factors. *Journal of the American Medical Association, 318*(12), 1180–1182. doi:10.1001/jama.2017.9087

Werner, C. & Engelhard, K. (2007). Pathophysiology of traumatic brain injury. *British Journal of Anaesthesia, 99*(1), 4-9. doi:10.1093/bja/aem131

ABOUT THE AUTHORS

Pictured: From left to right, Zizzo. Tankersley, and Bermingham smile with their Denver Metro Regional Science Fair awards. Their mentor, Dr. Link of CU Boulder, is not pictured.

Zizzo, Tankersley, and Bermingham are all juniors who are a part of the Rock Canyon High School Biotechnology Program. Their love of biotech began with the Introduction to Biotechnology course, where they learned rudimentary lab skills and gave them a glimpse into the astonishing advancements in the biotechnology field.

Through their experience in the Experimental Design in Biotechnology course, they discovered the challenges that working in a team poses as well as the benefits of overcoming adversity within research. The team learned how important communication is to research as they discussed ideas and worked through problems, such as developing a proficient research plan and learning extensive protocols. Each member learned the art of overcoming obstacles and troubleshooting when certain protocols did not work, teaching them focused resiliency along with stamina. Each student will be able to apply these lessons to her everyday life, as well as future research endeavors.

From a young age, Zizzo has always had a passion for science. She began fostering her love for science when she was in the fourth grade by volunteering at the Denver Museum of Nature and Science and sought out numerous other opportunities to learn more and teach others various scientific concepts. This Biotechnology Research course has helped further foster her love of science by providing her with the experience of designing, performing, and then presenting a research project. Zizzo's career goal is to advance medicine through research.

Tankersley is very passionate about science, and through this year has developed a love for researching especially in the Biotechnology field. She has learned a significant amount about how difficult the life of a researcher is through this Biotechnology Research course. This class has taught her the importance of being proactive in all aspects of research, and the importance of communication within a group. She has grown in her knowledge of science, becoming more excited and passionate to pursue a future career in research. She loved the challenging nature of this year's research course and hopes to carry her newfound respect for science into her daily life.

Bermingham's passion for science began after her mother went back to school to get her radiology technician degree during her third-grade year. Bermingham was fascinated with her biology and anatomy courses and was inspired by her mother's determination to start a career she loved. From this point on, Bermingham took as many science classes as she could and loved every single one of them. The Rock Canyon Biotechnology Program not only allowed her to do something she was passionate about but also gave her the resources she needed to develop perseverance and problem-solving skills that will benefit her for the rest of her life. This course allowed Bermingham's love of science to flourish and gave her the tools to succeed in her future science career.

The effect of traumatic brain injury on learning and memory in *Drosophila melanogaster* and the efficacy of piracetam as a TBI treatment

S. H. Naik, K. B. Dick, C. D. Peters, & S. M. Petri
Department of Science, Principles of Experimental Design in Biotechnology, Rock Canyon High School, Highlands Ranch, Colorado, USA

Traumatic brain injury (TBI) has increased at an alarming rate in the last decade yet few effective pharmaceutical treatments exist beyond anti-inflammatories, making research on potential TBI model organisms essential. In this research, we investigated how a TBI-inducing event affects learning and memory in *Drosophila melanogaster* and how subsequent treatment with piracetam, a "smart drug", post-TBI treatment affects their performance in a learning and memory assay. We hypothesized that TBI treatment would impair learning and memory in *D. melanogaster* and that recovery in piracetam would improve performance. *D. melanogaster* were exposed to a TBI-inducing event in a High Impact Trauma (HIT) machine after which their ability to negatively associate the presence of a wire grid with an electric shock was tested in a learning and memory assay. In our no-TBI treatment control, the difference between the proportion of flies that avoided the grid before and after the shock was statistically significant (p = 0.0000129), indicating that the learning and memory assay effectively quantifies learning. When exposed to a TBI-inducing event in the HIT machine, the performance of the flies was significantly lower than the no-TBI control, verifying that the TBI treatment significantly affected the flies' learning and memory (p = 0.001037). In addition, recovery for 48 hours resulted in improved performance when compared to TBI-treated group given no recovery (p = 0.00154). In contrast, when flies recovered with piracetam for 24 or 48 hours post TBI treatment, no statistically significant effect on learning was observed when compared to the no piracetam recovery controls (p = 0.386 and p = 0.033254 respectively). It is important to note that the lowered performance of the TBI-treated *D. melanogaster* with 48 hour recovery in piracetam was trending towards statistical significance. More research is necessary to determine the efficacy of piracetam as a treatment for TBI.

Traumatic brain injury (TBI) is a leading cause of death worldwide and is a rapidly growing problem (Masel & DeWitt, 2010; Blennow, Hardy, & Zetterberg, 2012). In the United States alone, 1.7 million people are diagnosed with TBI annually, excluding all undiagnosed cases; TBI accounted for 275,000 hospitalizations, 1,365,000 emergency visits, and 52,000 deaths between 2002 and 2006 (Faul, Xu, Wald, & Coronado, 2010). Concussions, a specific type of TBI, are on the rise, in youth aged 5 to 25 with an alarming 500% increase from 2010-2014 (Fox News Network, 2016). Due to this increasingly significant issue in youth, it is imperative that more research is conducted to find treatments and therapeutic agents. TBI can take months, years, or an entire lifetime to heal because of heterogeneity in the severity, symptoms, complications, and unpredictable mechanisms of each individual TBI case. Consequently, doctors are unable to accurately predict the healing time of TBI, leaving a vast need for more scientific research to understand the neurological mechanisms occurring post TBI and, thus, the recovery process (Johnson, 2010).

A key part of tackling the recovery process is understanding the mechanisms behind traumatic brain injuries. In closed head injuries, the type of TBI we investigated, an outside force induces direct damage to the brain as it impacts

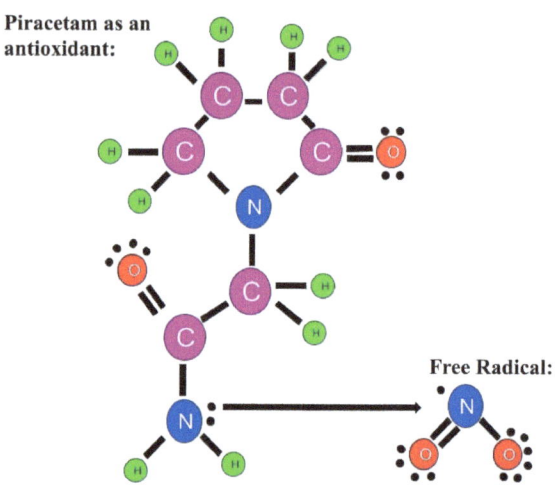

Picture 1: Free radicals have a lone electron which makes them very reactive. They take electrons from other molecules, destabilizing them and causing a chain reaction of destruction. Antioxidants donate extra electrons to stabilize the free radicals and stop the chain reaction.

the skull, resulting in broken blood vessels, swelling, and damaged neurons. The initial physical damage is referred to as the primary injury, while damage from the subsequent harmful chemical reactions are called secondary injuries.

Oxidative stress, a type of secondary injury, occurs when an excess of free radicals is released into the brain upsetting its chemical balance (Ansari, Roberts, & Scheff, 2008). Free radicals destabilize electrons in essential biomolecules, such as DNA, lipids, and proteins, causing a chain reaction, and potentially triggering apoptosis (Lewén, Matz, & Chan, 2009; Ahamed et al., 2010). These serious chemical imbalances caused by TBI have inspired many new research initiatives towards the discovery of treatments that target the damage caused by secondary injuries. One area of interesting new research is the efficacy of antioxidants to reduce oxidative stress as a future treatment of TBI (Pham-Huy, He, & Pham-Huy, 2008) (**Pic 1**). Another area of new research is investi-gating the acetylcholine (ACh) pathway, thought to be responsible for modulating learning and memory (Chen & D'Esposito, 2010). This research has shown promise in the investigation into new drug targets as TBI often causes chronic memory dysfunction (Girgis, Pace, Sweet, & Miller, 2006).

One drug with potential to treat TBI on a molecular level is piracetam (**Pic. 2**). Piracetam is a nootropic "smart drug" that regulates neurotransmitter systems essential to the function of the memory centers of the brain found in the hippocampus and cortex (Aigner, 1995; Winblad, 2006). Although not fully understood, scientific research suggests that piracetam affects membrane fluidity leading scientists to hypothesize it alters neuronal membranes allowing for improved communication between cells through ion transmission (Winblad, 2006). Studies also suggest piracetam improves ACh neurotransmission in aging mice by healing and increasing the number of nicotinic ACh postsynaptic receptors (nAChR) required for ACh

Picture 2: The piracetam used in our experiment came in the form of a white powder.

function (Pilch & Müller, 1988; Stoll, Schubert, & Müller, 1992). Improved neurotransmission of ACh is very likely responsible for the positive boost in learning and memory reported by people who take piracetam due to the essential role ACh plays in encoding memories, memory recall, and turning short-term into long-term memories (Micheau & Marighetto, 2011). Cholinergic treatments may have potential to treat TBI as a part of this system located in the hippocampus and frontal cortex is damaged (Shin & Dixon, 2015; Girgis et al., 2006).

Not only could piracetam act as a cholinergic treatment, but it also has potential in healing the brain, especially parts damaged from TBI. Studies conducted on rats with alcohol-damaged brains found piracetam has neuroprotective effects against neuronal loss and long-term change in the brain and is able to aid in neuroplasticity by increasing hippocampal synapses (Brandao, Cadete-Leite, Andrade, Madeira, & Paula-Barbosa, 1996; Brandao, Paula-Barbosa, & Cadete-Leite, 1995). Not only could piracetam lessen the damage

caused by TBI on learning and memory, it has the ability to heal cellular damage caused by secondary injuries. It has been found to act as an antioxidant *in vitro,* especially in high concentrations, ridding blood of poisonous free radicals and reducing oxidative stress (Horvath et al., 2002). Antioxidants give up extra electrons to stop the chain reaction of damage from free radicals before a fatal level of a cell's biomolecules are destabilized as well as prevent oxidative stress by stabilizing free radicals and transition metal radicals (Pham-Huy, He, & Pham-Huy, 2008; Young & Woodside, 2001). Antioxidant agents have also been shown to reduce the extent of brain swelling or edema in a concussion, potentially further reducing the harm of the TBI secondary injuries (Smayda, 2018). Numerous preclinical research studies demonstrate that treating TBI with antioxidant compounds could potentially be therapeutic; however, there are a lack of sufficient clinical results at this time (Petraglia, Winkler & Bailes, 2011). A study conducted in 2007 aimed to deter-mine if piracetam had an effect in alleviating symptoms of post-concussion syndrome, including difficulty with memory. It was found that piracetam had many positive effects in reducing symptoms as well as aiding blood flow and reducing perfusion abnormalities (Agrawal & Gowda, 2007). Piracetam is a very promising TBI treatment as many scientists have reported its ability to amplify the cholinergic system and reduce the severity of TBI symptoms.

In order to conduct preliminary research with piracetam and TBI, a model organism with similar pathways to humans is needed. *D. melanogaster* are an ideal model for this TBI research study for numerous reasons. Their nervous system physiology is similar to that of humans, and their memory centers, the mushrooms bodies, are homologous to the hippocampus in humans (Davis, 1993). The area of the brain containing the mushroom bodies swells during TBI, potentially causing similar memory problems to those in humans (**Pic. 3**). Additionally, *D. melanogaster* are nonvertebrate organisms, making them ideal models because they are unregulated by IACUC guidelines and are easy to perform drug screening tests on. They are small, inexpensive, and breed quickly allowing for the abundance of flies needed for experimentation. *D. melanogaster* can also model genetic variation in a population affecting the susceptibility of the human patient to neurodegeneration or death (Gavett, Stern, Cantu, Nowinski, & McKee, 2010; Gavett, Stern, & Mc-Kee,

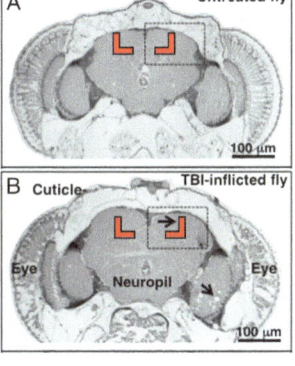

Picture 3: These are brain scans of a fly before and after TBI. Swelling is visible in the area of the brain containing the mushroom bodies, denoted in red. Image from "A *Drosophila* model of closed head traumatic brain injury," by R. J. Katzenberger, C. A. Loewen, D. R. Wassarman, A. J. Petersen, B. Ganetzky, & D. A. Wassarman, 2013, *Proceedings of the National Academy of Sciences of the United States of America, 110,* E4151-E4159. Copyright 2013 of Proceedings of the National Academy of Sciences of the United States of America. Adapted with permission.

2011). They are an ideal model organism for piracetam research because the drug has the potential to affect flies' learning as studies have shown ACh may be a key neurotransmitter in olfactory learning in *D. melanogaster* (Barnstedt *et al.*, 2016).

In order to measure learning and memory in the *D. melanogaster,* we engineered and tested a learning and memory assay in order to confirm that *D. melanogaster* could negatively associate the presence of a wire grid with a shock. We then built and tested a HIT machine used to inflict TBI on the *D. melanogaster.* After confirming that the HIT machine was able to produce differences in learning and memory in the TBI-treated *D. melanogaster*, we tested the effect of recovery with piracetam on their performance in the learning and memory assay.

Our learning and memory assay was modeled after a study that investigated the ability of *D. melanogaster* to learn to avoid a scent with a negative stimulus. In this study, approaching one of the scents resulted in the flies receiving a small shock while the other scent did not result in the negative stimulus. The flies' ability to learn to avoid the shock was found to be statistically significant, demonstrating the efficacy of a shock assay in measuring learning and memory in fruit flies (Quinn, Harris & Benzer, 1974). The HIT machine we constructed to induce TBI in the flies was originally designed and tested to determine how well *D. melanogaster* could model closed head TBI (Katzenberger *et al.*, 2013). The original HIT machine consisted of a plastic fly vial (HIT vial) on the end of a spring attached to a wooden board and polyurethane pad base (**Pic. 4**). The severity of TBI varied within the HIT vial, with sample groups evenly

split between sexes, modeling the variance in the severity of human primary injuries (Katzenberger *et al.*, 2013). They observed that the flies were momentarily paralyzed at the bottom of the vial directly after impact but regained movement within 5 minutes and did not sustain injuries to their appendages (**Pic. 4**). Neither the amount of flies in the HIT vial (10-60 flies), the length of the rest period between strikes (1-60 minutes), or the sex of the flies had a statistically significant effect on the mortality rate within 24 hours. They also determined *D. melanogaster* aged 0-3 and 4-6 days had a 24 hour mortality index of approximately 25% (Katzenberger *et al.*, 2013).

In our research, we investigated the effects of TBI on *D. melanogaster* learning and memory, and the efficacy of piracetam as a treatment for one main symptom of a TBI, impaired learning and memory. We hypothesized piracetam would improve the performance of *D. melanogaster* in the learning and memory assay post TBI-inducing treatment.

METHODS

In the first part of our experiment, we compared the learning of the *D. melanogaster* post TBI-inducing treatment to a no treatment control group. In the second part of the experiment, we observed the effects of piracetam treatment on the performance of the *D. melanogaster* in the learning and memory assay post TBI-inducing event.

D. melanogaster Care

Apterous *D. melanogaster* (PetSmart) were bred in 10.16 cm tall, 3.175 cm diameter plastic vials with 5.0 g of Formula 4-24 Instant *Drosophila* media (Carolina Biological) and 20 mL of water. They were housed at room temperature under ambient lighting conditions in the Biotechnology Research Lab. For all trials, *D. melanogaster* were between 0 and 6 days old.

Learning and Memory Assay

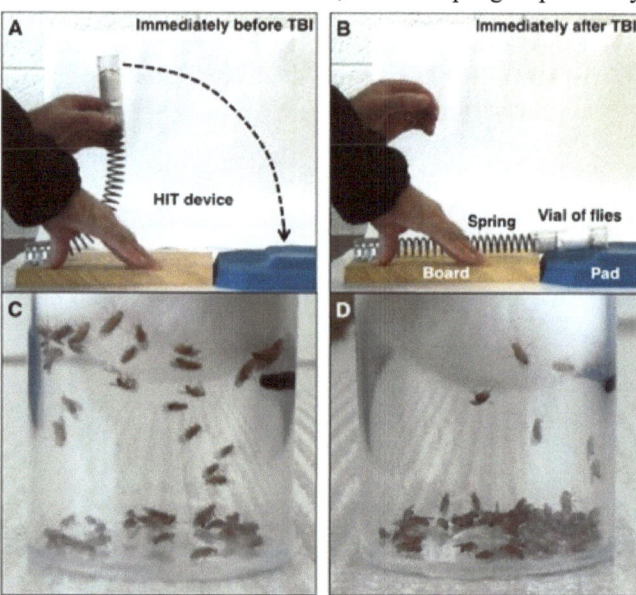

and the way the flies act immediately after being afflicted with TBI. **A)** The spring with a fly vial attach to it is lifted up to a 90° angle. **B)** The spring is then released and hits on a hard foam. **C)** This show the flies before being released acting normally. **D)** After the flies were released on to the foam pad the flies fell to the bottom of the vial and were temporarily paralyzed. Image from "A *Drosophila* model of closed head traumatic brain injury," by R. J. Katzenberger, C. A. Loewen, D. R. Wassarman, A. J. Petersen, B. Ganet-zky, & D. A. Wassarman, 2013, Proceedings of the National Academy of Sciences of the United States of America, 110, E4151-E4159. Copyright 2013 of Proceedings of the National Academy of Sciences of the United States of America. Reprinted with permission.

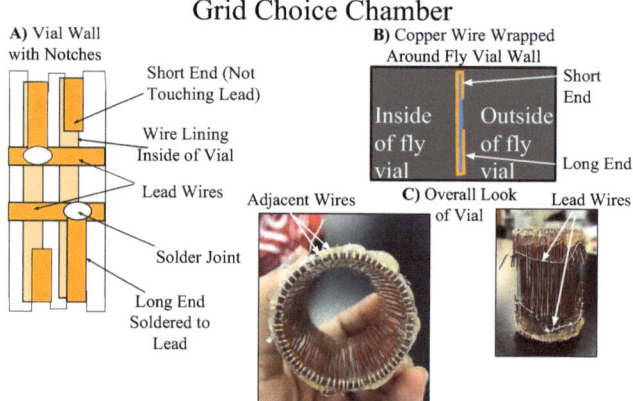

Figure 1: These are diagrams and images of our grid choice chamber. **A)** This is an outside close-up of our chamber design. There were notches cut out of the top and bottom of the plastic vial. The copper wires were wrapped around the inside and secured by the notches. The wire on the right was longer around the top of the vial, and the left was longer around the bottom, alternating around the whole vial. The long ends on the top were soldered to one lead wire and the long ends on the bottom were soldered to the other lead wire. **B)** This is a cross-section demonstrating how we wrapped the wires around the vial. The wire spanned the entire vial vertically on the inside and had a long and short end on the outside. **C)** These are images of our completed grid choice chamber. No adjacent wires were touching. The two leads are visible on the right picture.

We used the learning and memory assay to quantify the learning proficiency of a group of *D. melanogaster* by counting the number of *D. melanogaster* that learned to avoid a grid based on an electric shock (Quinn *et al.*, 1974). In order to build the learning and memory assay, we created a rest chamber, clear choice chamber, and a grid choice chamber. The rest chamber was a 10.16 cm tall, 3.175 cm diameter plastic fly vial with both ends open. The clear choice chamber was a 5.6 cm tall 3.175 cm diameter plastic fly vial with both ends open, and the grid choice chamber was built around a 5.6 cm tall, 3.175 cm diameter plastic fly vial. The grid choice chamber was built with 66 individual 22 gauge copper wires wrapped vertically around the vial at intervals of 1 mm apart. Each copper wire spanned the entire height of the inside surface of the vial and was held in place by notches in the plastic (**Fig. 1**). We attached a positive and negative lead to every other wire so no two adjacent wires touched. To administer the shock, two 9V batteries were connected to the grid choice chamber and a switch with alligator clips (**Fig. 2**).

Shocking Device

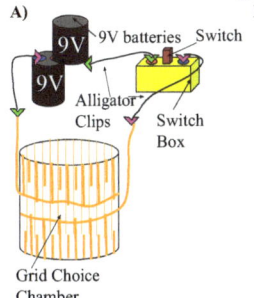 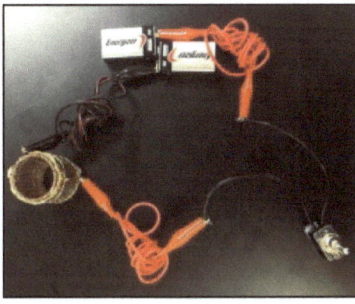

Figure 2: This diagram and image depict our shocking device. **A)** An open circuit was created by attaching two 9 volt batteries and a switch box to the lead wires on the grid choice chamber with alligator clips. The flies were placed inside the grid choice chamber and both ends were stop-pered. The flies closed the circuit by stepping on two adjacent wires and received a shock when the switch was turned on. **B)** This image depicts our fully assembled shocking device as it was used in the experiment.

To verify the efficacy of the learning and memory assay, we tested it with 54 apterous flies that had not received TBI treatment. The learning and memory assay consisted of four stages. First, the rest chamber was attached to the clear choice chamber with parafilm. We then inserted a foam stopper into the rest chamber 27 mm from the top. The *D. melanogaster* were inserted through the top of the clear choice chamber into the learning and memory assay, and the top was capped by a second foam stopper (**Fig. 3**). The flies were tapped onto the bottom stopper and the learning and memory assay was set upright, allowing the *D. melanogaster* one minute to decide whether to remain in the rest chamber or climb up into the clear choice chamber. In this stage, we observed the flies for normal movement and willingness to approach the upper chamber. In the second stage, the rest chamber was removed from the clear choice chamber. It was then attached (parafilm) to the grid choice chamber. Flies were again allowed one minute to choose whether to climb into the grid choice chamber or remain in the rest chamber (**Fig. 3**). We then counted the number of *D. melanogaster*

Learning and Memory Assay Grid Vial Set-up

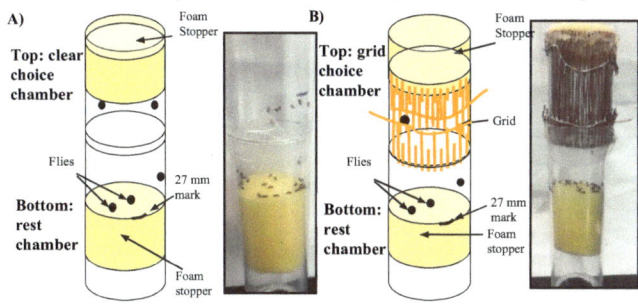

Figure 3: This depicts how we set up our learning and memory assay. **A)** For stage one, Flies were placed inside the assay and both ends were capped with foam stoppers. They were given a minute to choose whether to climb up the walls and sit in the top half (the clear choice chamber) or stay down in the rest chamber. **B)** For stage two, they were placed in the assay with the grid choice chamber attached on top, and both ends were stoppered. They were given a minute to choose whether or not to climb the walls and stand on the grid. No shock was administered in stage two. For stage three, the bottom stopper was pushed up to meet the grid choice chamber, ensuring the flies could not leave the grid choice chamber. A five second shock was administered. After their recovery time, the flies were placed back in the assay as depicted above, and they were given a minute to choose whether or not to approach the grid choice chamber.

that stayed in the rest chamber, to determine the number of flies that were naturally inclined to avoid the grid. For stage three, we attached the shocking device to the lead wires of the grid choice chamber (**Fig. 2**). All the *D. melanogaster* were tapped into the upper half of the assay and the bottom stopper was pushed all the way up to the grid choice chamber. A five-second shock was administered. Following this shock, the *D. melanogaster* were given between five and ten minutes to recover in an empty fly culture vial. In the fourth and final stage, the *D. melanogaster* were moved back to the rest chamber that was still attached to the grid choice chamber. They were given one minute to choose whether to climb into the grid choice chamber or remain in the rest chamber. In this stage, we observed the number of flies that avoided the grid now that it was negatively associated with a shock. The number of *D. melanogaster* that remained in the rest chamber during each stage was video recorded. Verifying the efficacy of this learning and memory assay was a critical first step before we were able to proceed to our experimental trials.

HIT Machine

We built the HIT machine to inflict TBI in the *D. melanogaster* based on the design in prior research (Katzenberger, *et al.*, 2013). The HIT machine was built by attaching a small plank of wood (39.37 cm x 5.08 cm x 7.62 cm) to a pad of hard foam (26.67 cm x 22.86 cm x 7.62 cm) and attaching a 35.56 cm long spring with a 2.54 cm diameter to the end of the board with a metal washer and screw (**Fig. 4**). We secured a fly vial, the HIT vial, to the end of the spring using duct tape. The *D. melanogaster* were placed in the HIT vial and a stopper was pushed in until there was a 2.54 cm gap from the bottom. To induce TBI, we lifted the spring with the HIT vial containing the *D. melanogaster* to a 90° angle and released it allowing the HIT vial to be slammed onto the pad of hard foam. We determined that striking the HIT vial containing

HIT Machine

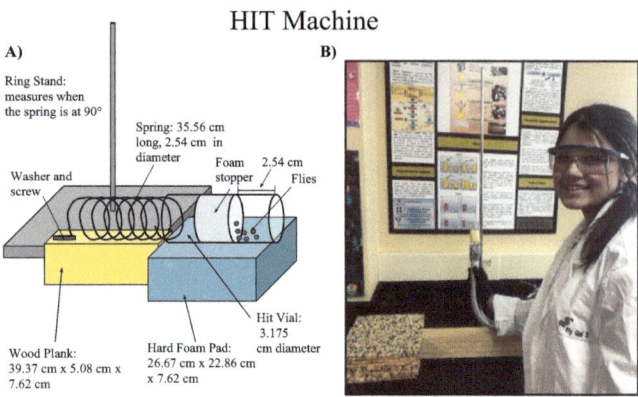

Figure 4: This image and diagram depict our HIT machine **A)** Flies were placed in the HIT vial and a foam stopper was pushed into the vial until it was 2.54 cm above the bottom of the vial. The HIT vial was attached to the spring with duct tape. The spring was pulled up until parallel with the ring stand (90° from the wood board). When let go, the HIT vial struck the foam pad, inflicting TBI in the flies. They were hit four times with five minute rest periods between strikes. **B)** Naik is demonstrating our HIT machine set up.

approximately 60 *D. melanogaster* four times with five minute rest intervals between each strike resulted in a mortality rate between 25-50%. This mortality rate ensured that the TBI treatment from the HIT machine was likely causing injury to the brain while not raising the mortality beyond reasonable limits for the purpose of having flies to conduct the experiment.

Mobility Assay
After *D. melanogaster* received the TBI treatment in the HIT machine, they were evaluated for successful mobility before being tested in the learning and memory assay (**Pic. 5**). We emptied the flies onto a large piece of butcher paper laid on a flat surface to examine them. Those with obvious physical injuries were removed. We then selected only flies that were moving unimpaired from the center, removing any flies that appeared injured, damaged, or behaving abnormally. Only unimpaired flies were put into the learning and memory assay and tested.

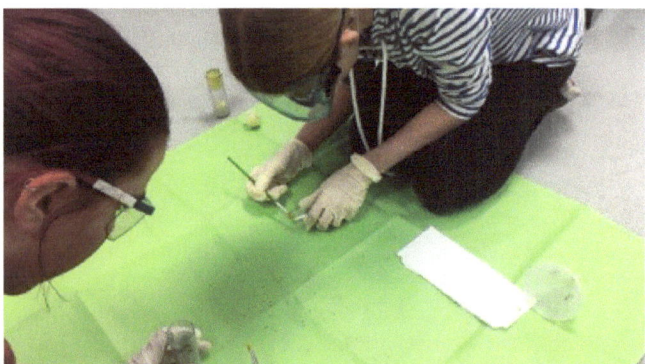

Picture 5: Peters and Dick perform the mobility assay. The flies were placed on the butcher paper and were observed for normal movement. Mobile flies were tested for learning, injured flies were removed.

Experiment One
In our first experiment, we tested if *D. melanogaster* that received TBI treatment in the HIT machine exhibited a lowered performance in the learning and memory assay as compared to the no-TBI treatment control. Before testing learning proficiency, approximately 60 apterous *D. melanogaster* were given TBI treatment in the HIT machine and were immediately assayed with the aforementioned mobility assay. We then tested the flies in the learning and memory assay. This experiment determined if exposure to a TBI-inducing event in *D. melanogaster* resulted in measurable changes in learning and memory.

Experiment Two
To determine the potential of piracetam as a treatment for the TBI symptoms of learning and memory dysfunction, TBI-treated *D. melanogaster* recovered for 24 or 48 hours in media infused with piracetam. To make the media, 20 mL of 0.1M piracetam was mixed with 5.0 g of media. The molarity of the piracetam was determined during pre-trials as the highest amount of piracetam that could be administered without inducing mortality. A control group that received TBI treatment but no piracetam was also tested for each recovery period. After the designated recovery period, the flies were assessed in the mobility assay after which their performance in the learning and memory assay was evaluated.

Data Analysis
In the learning and memory assay, to verify that the flies learned to avoid a wire grid by negatively associating it with an electric shock, we compared the proportion of flies that avoided that the grid before and after the shock using a one-tailed two proportion z-test with a significance level of $p = 0.05$. When analyzing the performance of the TBI-treated flies with no recovery as well as with recovery (with and without piracetam), we calculated the change in grid avoidance for each group by subtracting the proportion of flies that avoided the grid before the shock from the proportion that avoided it after the shock. The data was normalized to account for mortality during the assay. We then performed a two-tailed two proportion z-test on the data with a level of significance of $p = 0.025$.

RESULTS
In this investigation, we tested the efficacy of both the HIT machine and the learning and memory assay as well as the effects of treatment with piracetam on the performance of the *D. melanogaster* in the learning and memory assay after a 24 or 48 hour recovery time.

Experiment One
The data showed that the learning and memory assay was an effective tool to measure learning in *D. melanogaster*. The proportion of *D. melanogaster* that avoided the grid choice chamber before the shock was 24/54 and after the shock increased by 38.9% to 45/54. Using a one-tailed two proportion z-test, this difference was statistically significant ($p = 0.0000129$). Our data also confirmed that TBI treatment using the HIT machine negatively impacted the flies' ability to perform in the learning and memory assay. The 13.9% change in grid avoidance of the TBI-treated *D. melanogaster* with no recovery time was compared to the 38.9% change in

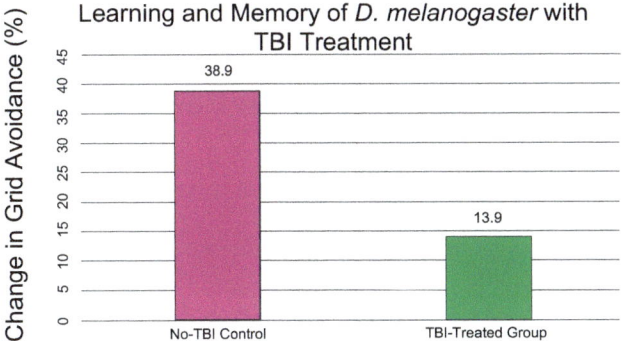

Graph 1: Change in grid avoidance was calculated by subtracting the proportion of flies that avoided the grid before the shock from the proportion that avoided it after the shock. In the no-TBI control group, the proportion of flies that avoided the grid before the shock was statistically different than the proportion that avoided it after the shock when compared with a one-tailed two proportion z-test (p = 0.0000129). There was also a statistically significant difference between the change in grid avoidances for the two groups displayed above when compared with a two-tailed two proportion z test (p = 0.001037).

grid avoidance of the no-TBI treatment control group (**Graph 1**). This difference was statistically significant when compared using a two-tailed two proportion z-test (p = 0.001037).

Experiment Two

After verifying the efficacy of both the HIT machine and the learning and memory assay and showing that the HIT machine affected learning, we moved onto trials with recovery and piracetam. *D. melanogaster* that were given 24 hours of recovery both with and without piracetam exhibited a change in grid avoidance (25.7% and 17.6% respectively) that was not statistically significant (two-tailed two proportion z-test) when compared to the 13.9% change in grid avoidance of the TBI-treated flies given no recovery time (p = 098739and p = 0.616 respectively). However, the *D. melanogaster* given 48 hours to recover without piracetam achieved a 41.2% change in grid avoidance which was statistically significant (two-

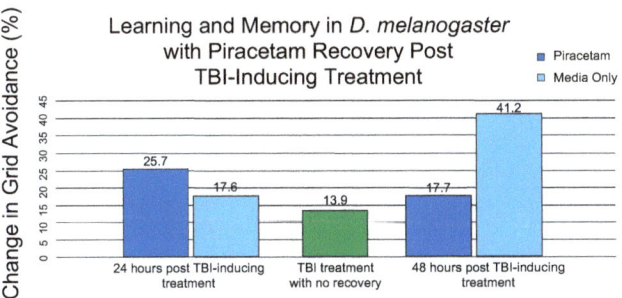

Graph 2: The change in grid avoidance for each blue group above was compared to the change in grid avoidance of the green group using a two-tailed two proportion z-test. The difference was not statistically significant for either group of flies given a 24 hour recovery, nor for the group given a 48 hour recovery time with piracetam (p = 0.616, 0.098739, and 0.616 respectively). The difference was statistically significant (p = 0.00154) for the group given a 48 hour recovery with no piracetam. The two groups given a 24 hour recovery were compared with a two-tailed two proportion z-test, but the difference was not significant with a p-value of 0.386. The difference between the 48 hour recovery groups was trending toward statistical significance when compared with a two-tailed two proportion z-test (p = 0.033254).

tailed two proportion z-test) when compared to the 13.9% change in grid avoidance of the TBI-treated flies given no recovery time (p = 0.00154). Interestingly, the group of flies given the same 48 hour recovery time but with piracetam only achieved a 17.7% change in grid avoidance which was not statistically different from the 13.9% change in grid avoidance of TBI-treated *D. melanogaster* with no recovery (p = 0.616) (**Graph 2**).

Next, we compared the change in grid avoidance of *D. melanogaster* with 24 hours and 48 hours recovery in piracetam to TBI-treated flies that recovered without piracetam for the same amount of time to determine if piracetam had a significant effect on the performance of the *D. melanogaster* in the learning and memory assay. The difference between the change in grid avoidance of the flies given a 24 hour recovery with piracetam (25.7%) and the flies given a 24 hour recovery without piracetam (17.6%) was not statistically significant when compared with a two-tailed two proportion z-test (p = 0.386). There was also no statistical difference between the change in grid avoidance of the flies given 48 hours to recover with piracetam (17.7%) or without piracetam (42.1%) with p-value of 0.033254 (two-tailed two proportion z-test). However, this difference was trending towards significance.

DISCUSSION

As a result of the vast growth in the prevalence of TBI, we felt it was necessary to conduct an experiment that targeted the secondary injuries caused by a TBI and its subsequent effects on learning in memory. In this investigation, we tested the effect of TBI on learning and memory in *D. melanogaster* and the efficacy of piracetam as a treatment. We hypothesized that receiving TBI treatment would significantly decrease learning and memory in *D. melanogaster* and piracetam would improve their performance in the learning and memory assay. To test this hypothesis, we built a HIT machine to induce TBI trauma in *D. melanogaster* and designed a learning and memory assay to quantify their ability to negatively associate a wire grid with an electric shock. Both the HIT machine and the learning and memory assay were shown to be effective. When we compared the difference between the proportion of flies that avoided the grid in the learning and memory assay before receiving the shock to the proportion that avoided the grid after the shock with a one-tailed two proportion z-test, it was statistically significant (p = 0.0000129). This indicates that the *D. melanogaster* were able to recognize that the wire grid would result in a shock and subsequently avoid it. In addition, we found that with TBI treatment in the HIT machine, the *D. melanogaster* exhibited impaired learning as compared to the no-TBI treatment control with a change in grid avoidance of 13.9% and 38.9% respectively (**Graph 1**). This data was statistically significant (p = 0.001037) when analyzed using a two-tailed two proportion z-test. This demonstrated the efficacy of our HIT machine, as well as supported our hypothesis that TBI treatment in *D. melanogaster* would result in lowered performance in the learning and memory assay.

When *D. melanogaster* were allowed to recover for 24 hours in regular media, there was no statistically significant improvement in their performance in the learning and memory assay (p = 0.616). It appeared that 24 hours of recovery did not make a difference and that the learning of the *D. melanogaster* remained significantly impaired post TBI treatment. A study was conducted on the effect of mild repetitive TBI on *D. melanogaster*, and it found that this disrupts the flies' ability to stay asleep, a common symptom in human TBI (Ayeh *et al.*, 2016). Other studies have demonstrated that in *D. melanogaster*, similar to humans, a lack of sleep harms their subsequent ability to react to a stimulus (Cirelli *et al.*, 2005). If the flies' sleep was impaired by the TBI-inducing event, their ability to respond to the shock in our learning and memory assay could also have been impaired. Inability to sleep regularly could explain why we did not observe improvement in their performance in the learning and memory assay at 24 hours post TBI treatment. However, it is also possible that a 24 hour recovery period is not enough time to restore the learning and memory capabilities of the *D. melanogaster* after this level of trauma is inflicted. When given 48 hours of recovery in normal media, there was a 41.2% change in grid avoidance of *D. melanogaster* which was 27.3 percentage points higher than the 13.9% change in grid avoidance of the TBI-treated flies with no recovery time (**Graph 2**). While more research is needed to determine a recovery time for TBI in *D. melanogaster*, this statistically significant recovery (p = 0.00154) shows promise that the *D. melanogaster* are able to recover from TBI as measured by the learning and memory assay, and could be useful in future research regarding recovery times and mechanisms in more complex organisms.

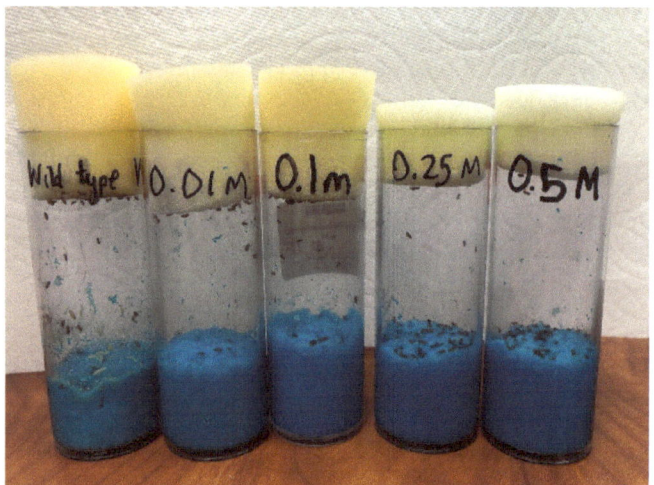

Picture 6: This image several different dosages of piracetam. All dosages 0.25M and higher were fatal. A 0.1M dosage was used in the experiment.

When *D. melanogaster* were given 24 hours to recover in media infused with 0.1 M piracetam (**Picture 6**) after receiving TBI-inducing treatment, a 25.7% change in grid avoidance was observed, however, it was not statistically significant when compared to the TBI treated *D. melanogaster* with no recovery time (p = 0.098739). Interestingly, the *D. melanogaster* that recovered for 48 hours in piracetam showed a lowered performance in learning and memory in comparison to the 48 hour no

piracetam group, by 23.5 percentage points. Though not statistically significant (p = 0.033254), these results were trending toward significance, suggesting that piracetam treatment not only did not aid in recovery from the TBI treatment, but that it may have been detrimental. Prev-ious research has shown piracetam to increase the length of paradoxical sleep in rats (Wetzel, 1985). If the flies' sleep cycle was unsettled by their injury, piracetam had the potential to mitigate the problem. As our results contradict this hypothesis, it is possible that our issue was in the piracetam dosage that was administered to the *D. melanogaster*, as it was the highest dosage we could administer without causing death. As the majority of things in extreme amounts tend to cause harm, it is important to repeat the experiment with more trials and further research on different doses and exposure periods to piracetam to concretely determine its effect on learning in *D. melanogaster*.

An important application of our research was the efficacy of the combination of the HIT machine for inflicting TBI symptoms and the learning and memory assay for accurately measuring these symptoms. The success of *D. melanogaster* in prior TBI research illustrates their value as a model organism. The development and validation of these two assays will allow for further research regarding TBI and its effects on learning and memory in *D. melanogaster,* which is crucial for developing a better understanding of TBI symptoms and potential treatments. Future steps for this research include increasing the number of trials and extending the length of recovery and exposure and lowering the dosage of piracetam, in order to definitively determine its effects on learning in *D. melanogaster*.

For our experiment, one potential source of error occurred during the mobility assay. We were reliant on sight to determine the ability of *D. melanogaster* to move normally. There was probably an inevitable margin of human error through the mobility assay that could have resulted in a higher learning percentage than would be correct as immobile flies would have stayed at the bottom of the assay unable to move. An-other source of error is that we could not control or measure the amount of piracetam that was consumed by each individual fly in the group, however, we accounted for this discrepancy by assessing the *D. melanogaster* as a population rather than on an individual basis.

ACKNOWLEDGMENTS
We would like to thank the following people for supporting and donating to our project. Thank you to Shawndra Fordham, our experimental design teacher at Rock Canyon High School and mentor, for her time, help, and equipment that allowed this project to be possible. We would also like to thank our mentor Brenna Dennison, a graduate student at Colorado University Anschutz medical campus, for all her help and support during the process of our research. We would like to thank the Rock Canyon High School science department and Nayan Naik for funding our project without whom we would not have been able to conduct our research. We would also like to thank Bryan Winkelman, the Rock Canyon High School teacher librarian, for his help, technical support of our project, and for running the biotech research

website. Additionally, we would like to thank Gwendolyn Karaba for her advice with statistical analysis without whom we would never have been able to determine the statistical significance of our data. We would like to thank David Ferguson for offering his help and expertise with the chemical aspects of our project and for ensuring our safety throughout our research. We would like to thank Kyler Barker, Daniel Jibson, and Giulio Cesarone for their help with the engineering phase of our project as we designed and constructed the electrical grid for the grid choice chamber of the learning and memory assay. We would also like to thank Jeffrey Seaquist for his generous donation of the *D. melanogaster* media used in our project. Thank you to Scott Kroman, Darrell Yashu, and Randy Lee, employees at Home Depot, for their work to help us find materials to build our assays and for their generous donation of the foam padding and some other materials used in our HIT machine. We would like to thank Nick and Karen Peters, Nathan and Lisa Dick, and Nayan and Purvi Naik for their generous donations to fund our project in the preliminary stages. Without their support, we would never have completed our research in the time allotted. Additionally, we would like to thank Megan Hupka, Tabitha Samuel, and Seth Rohr for their donation of piracetam. Finally, we would like to thank Rock Canyon High School and Douglas County School District for providing space and equipment to perform our experiment.

REFERENCES

Agrawal, D., & Gowda, N. K. (2007). Piracetam in postconcussion syndrome: preliminary results of a randomized study using SPECT. *Indian Journal of Neurotrauma*, 4(2):109-114. doi: 10.1016/S0973-0508(07)80024-4

Ahamed, M., Posgai R., Gorey, T. J., Nielsen, M., Hussain, S. M., & Rowe, J. J. (2010). Silver nanoparticles induced heat shock protein 70, oxidative stress and apoptosis in *Drosophila melanogaster*. *Toxicology and Applied Pharmacology*, 242(3):263-269. doi: 10.1016/j.taap.2009.10.016

Aigner, T. G. (1995). Pharmacology of memory: cholinergic—glutamatergic interactions. *Current Opinion in Neurobiology*, 5(2):155-160. doi: 10.1016/0959-4388(95)80021-2

Ansari, M. A., Roberts, K. N., & Scheff, S. W. (2008). Oxidative stress and modification of synaptic proteins in hippocampus after traumatic brain injury. *Free Radical Biology and Medicine*, 45(4):443-452. doi: 10.1016/j.freeradbiomed.2008.04.038

Ayeh, B., Arysa , G., Ruth, E., Roxanne, W. K., Brandon, M., Nadja, E.,... Eric, P. R. (2016). Using *Drosophila* as an integrated model to study mild repetitive traumatic brain injury. *Science Reports*, 6: 25252. doi: 10.1038/srep25252

Barnstedt, O., Owald, D., Felsenberg, J., Brain, R., Moszynski, J. P., Talbot, C. B., ...Waddell, S. (2016). Memory-Relevant Mushroom Body Output Synapses Are Cholinergic. *Neuron*, 89(6):1237-1247. doi: 10.1016/j.neuron.2016.02.015

Blennow, K., Hardy, J., & Zetterberg, H. (2012). The neuropathology and neurobiology of traumatic brain injury. *Neuron*, 76(5):886–899. doi: 10.1016/ j.neuron.2012.11.021

Brandao, F., Cadete-Leite, A., Andrade, J.P., Madeira, M.D., & Paula-Barbosa, M. M. (1996). Piracetam promotes mossy fiber synaptic reorganization in rats withdrawn from alcohol. *Alcohol*, 13(3):239–249. doi: 10.1016/0741-8329(95)02050-0

Brandao, F., Paula-Barbosa, M. M., & Cadete-Leite, A. (1995). Piracetam impedes hippocampal neuronal loss during with-drawal after chronic alcohol intake. *Alcohol*, 12(3):279–288. doi: 10.1016/0741-8329(94)00107-O

Cirelli, C., Bushey, D., Hill, S., Huber, R., Kreber, R., Ganetzky, B., & Tononi, G. (2005). Reduce sleep in *Drosophila Shaker* mutants. *Nature International Journal of Science*, doi: 10.1038/nature03486

Chen, A. J. W., & D'Esposito, M. (2010). Traumatic Brain Injury: From Bench to Bedside to Society. *Neuroview*, 66(1):11-14. doi: 10.1016/j.neuron.2010.04.004

Davis, R. L. (1993). Mushroom bodies and *Drosophila* learning. *Neuron*, 11(1):1-14. doi: 10.1016/0896-6273(93)90266-T

Faul, M., Xu, L., Wald, M. M., & Coronado, V. G. (2010, March) Traumatic Brain Injury in the United States: Emergency Department Visits, Hospitalizations and Deaths 2002-2006. *CDC Stacks Public Health Publications*. Retrieved from https://stacks.cdc.gov/view/cdc/5571

Fox News Network. (2016, July 11). Data suggest increase in reported youth concussions; upwards of 500 percent since 2010. Retrieved from http://www.foxnews.com/health/2016/07/11/data-suggest-increase-in-reported-youth-concussions-upwards-500-percent-since-2010.html

Gavett, B. E., Stern, R. A., Cantu, R. C., Nowinski, C. J., & McKee A. C. (2010). Mild traumatic brain injury: A risk factor for neurodegeneration. *Alzheimer's Research & Therapy*, 2(3):18. doi: 10.1186/alzrt42

Gavett, B. E., Stern, R. A., & McKee, A. C. (2011). Chronic traumatic encephalopathy: A potential late effect of sport-related concussive and subconcussive head trauma. *Clinics in Sports Medicine*, 30(1): 179–188, xi. doi: 10.1016/j.csm.2010.09.007

Girgis, F., Pace, J., Sweet, J., & Miller, J. P. (2006). Hippocampal Neurophysiologic Changes after Mild Traumatic Brain Injury and Potential Neuromodulation Treatment Approaches. *Frontiers in Systems of Neuroscience*, 10(8). doi: 10.3389/fnsys.2016.00008

Horvath, B., Marton, Z., Halmosi, R., Alexy, T., Szapary, L., Vekasi, J., ...Toth, K. (2002). In vitro antioxidant properties of pentoxifylline, piracetam, and vinpocetine. *Clinical Neuropharmacology*, 25(1):37-42. doi: 10.1097/00002826-200201000-00007

Johnson, G. (2010). Traumatic Brain Injury Survival Guide. Retrieved from http://www.tbiguide.com/getbetter.html

Katzenberger, R. J., Loewen, C. A., Wassarman, D. R., Petersen, A. J., Ganetzky, B., & Wassarman, D. A. (2013) A *Drosophila* model of closed head traumatic brain injury. *Proceedings of the National Academy of Sciences of the United States of America*, 110(44), E4151-E4159. doi: 10.1073/pnas.1316895110

Lewén, A., Matz, P., & Chan, P. H. (2009). Free radical pathways in CNS Injury. *Journal of Neurotrauma*, 17(10). doi: 10.1089/neu.2000.17.871

Masel, B. E., & DeWitt, D.S. (2010). Traumatic brain injury: A disease process, not an event. *Journal of Neurotrauma*, 27(8):1529–1540. doi: 10.1089/neu.2010.1358

Micheau, J., & Marighetto, A. (2011). Acetylcholine and memory: A long, complex and chaotic but still living relationship. *Behavioral Brain Research*, 221(2):424-429. doi: 10.1016/j.bbr.2010.11.052

Petraglia, A., Bailes, J., & Winkler, E. (2011). Stuck at the bench: Potential natural neuroprotective compounds for concussion. *Surgical Neurology International*, 2(1), 146. doi: 10.4103/2152-7806.85987

Pham-Huy, L. A., He, H., & Pham-Huy, C. (2008). Free Radicals, Antioxidants in Disease and Health. *International Journal of Biomedical Science*, 4(2):89-96. doi: 10.4172/0974-8369.1000214

Pilch, H., & Müller, W. E. (1988). Piracetam elevates muscarinic cholinergic receptor density in the frontal cortex of aged but not of young mice. *Psychopharmacology*, 94(1):74–78. doi: 10.1007/BF00735884

Quinn, W. G., Harris, W. A., & Benzer, S. (1974). Conditioned Behavior in *Drosophila melanogaster*. *Proceedings of the National Academy of Sciences of the United States of America*, 71(3):708-712. doi: 10.1073/pnas.71.3.708

Shin, S. S., & Dixon, C. E. (2015). Alterations in Cholinergic Pathways and Therapeutic Strategies Targeting Cholinergic System after Traumatic Brain Injury. *Journal of Neurotrauma*, 32(19):1429-1440. doi: 10.1089/neu.2014.3445

Smayda, R. (2018). What happens to the brain during a concussion? *Scientific American*. Retrieved from https://www.scientificamerican.com/article/what-happens-to-the-brain/

Stoll, L., Schubert, T., & Müller, W. E. (1992). Age-related deficits of central muscarinic cholinergic receptor function in the mouse: Partial restoration by chronic piracetam treatment. *Neurobiology of Aging*, 13(1):39–44. doi: 10.1016/0197-4580(92)90006-J

Wetzel, W. (1985). Effects of nootropic drugs on the sleep-waking pattern of the rat. *Biomedica Biochimica Acta*, 49(5):405-11. Retrieved from https://europepmc.org/abstract/med/3936493

Winblad, B. (2006). Piracetam: A Review of Pharmacological Properties and Clinical Uses. *Central Nervous System Drug Reviews, 11*(2):169-182. doi: 10.1111/j.1527-3458.2005.tb00268.x

Young, I. S., & Woodside, J. V. (2001). Antioxidants in health and disease. *Journal of Clinical Pathology, 54*(3):176–186. doi: 10.1136/jcp.54.3.176

ABOUT THE AUTHORS

Pictured left to right: Naik, Peters, Dick, and their mentor, Brenna Dennison

Naik is currently a junior at Rock Canyon High School. Throughout her three years in high school, she has taken numerous courses in math and science including AP Calculus and AP Chemistry. The rise in incidents of high school traumatic brain injury, specifically concussions, in conjunction with the experiences of her partners, influenced Naik's passion for research regarding the effect of TBI on learning and memory. For her senior year, Naik plans to serve as a teacher assistant and mentor for the Rock Canyon High School Biotechnology Program. The lab skills and the soft skills that Naik has learned from the program have better prepared her to succeed in her prospective field of molecular biology after her graduation.

Peters is currently a senior and has taken Biology, Chemistry, Foundation of Physics, Introduction to Biotechnology, Experimental Design in Biotechnology and other science courses during her high school career. When Peters was young, she fell down the stairs and received a concussion. This experience led her to become interested in TBI research. During the time Peters has spent in the biotechnology program and Experimental Design in Biotechnology course, she has developed advanced lab skills. This experience has set her up very well to continue her education at Rutgers University- New Brunswick.

Dick has taken Honors Biology, Foundations of Physics, and Honors and AP Chemistry science courses throughout her high school career, giving her a wide range of scientific understanding. When she was 10 years old, she fell off a backyard zipline and got a concussion, inspiring her interest in TBI research. She has benefitted and grown tremendously from the Rock Canyon High School Biotechnology Program and the Experimental Design in Biotechnology course, learning high level lab skills and technical writing. This experience has fostered her love of science and has prepared her very well to go into a Biochemistry major next year at the University of Toronto.

Naik, Peters, and Dick have gained an immense number of technical and personal skills from the Rock Canyon High School Biotechnology Program and Experimental Design.

During the course of their research, they learned how to be creative, thorough, and resilient in their thinking when faced with difficult questions and seemingly insurmountable obstacles.

Danio rerio embryos as a model organism of secondary injury mechanisms in traumatic brain injury research

K. M. Zilligen & S. M. Petri
Department of Science, Principles of Experimental Design in Biotechnology, Rock Canyon High School, Highlands Ranch, Colorado, USA

Traumatic brain injury (TBI) often results in death and disability and are dramatically increasing in occurrence. It is important to understand and discover new ways to aid in the treatment of TBI. The purpose of this investigation was to explore *Danio rerio* (zebrafish) embryos as potential model organisms for future TBI research. Embryos are considered vertebrates 7 days post fertilization (dpf) making them an ideal candidate for early-stage research to identify drug targets and better understand the mechanisms of TBI. To induce TBI, 3-4 dpf embryos were placed in a shaking incubator at 250 rpm for 1 minute. A mortality, response to touch, and head size assay were measured to identify symptoms of TBI. To eliminate the variable of innate size difference between embryos, the body length was proportionalized to the length of the fore, mid, and hindbrain. The TBI induction resulted in a difference between the mortality assay control and experimental groups. The touch and head size assays were shown to have no significant difference between the control and experimental groups. These findings do not support the hypothesis that zebrafish embryos can be used as effective model organism of TBI via treatment in a shaking incubator as used in this project. Although the embryos in this research failed to show TBI symptoms of head edema and delayed response time, damage was inflicted. Therefore, further research should explore zebrafish embryos as potential model organism of TBI when induced through optimized methods other than a shaking incubator.

Traumatic brain injuries (TBI) are the result of impact severe enough to produce violent movement of the brain (Prins, Greco, Alexander, & Giza, 2013). Often such impact occurs in events such as vehicle accidents or falls (**Fig. 1**) (Taylor, Bell, Breiding, & Xu, 2017). Within TBI, the initial impact is categorized as the primary injury whereas the following symptoms are known as the secondary injury: memory loss, sensitivity to light and sound, loss of consciousness, chronic headaches and even death (Murthy, Bhatia, Sandhu, Prabhakar, & Gogna, 2005). Additional physical symptoms include brain edema, hypotension, bruising, and bleeding, causing a lack of oxygen to reach the brain by hampering its ability to travel, furthering damage (Murthy, *et al.*, 2005). TBI is known to reduce processing speed and basic cognitive function withal by weakening neuronal membranes, allowing for the release of free radicals into extracellular space (P. Thompson, personal communication, August 22, 2018). Another hindersome quality found in the secondary injuries of TBI is visible damage to the axon of neurons

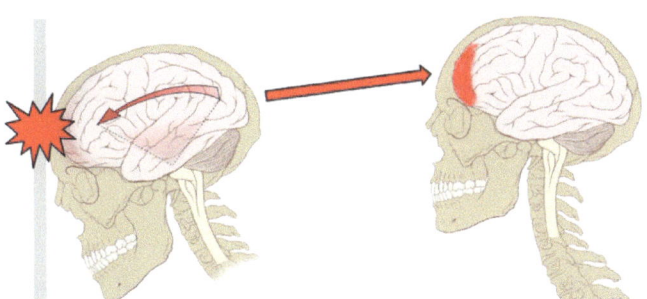

Figure 1: Movement of the brain within the head resulting in impact causing damage. From "Concussion Anatomy.png," by M. Andrews, 2012, CC-BY-SA-3.0.

(Katz, Cohen, & Alexander, 2015). Damage to this area is significant as it is responsible for intercommunication of neurons and thus results in changes in cognition, memory, and behavior (Rudy, Maia, & Kutz, 2016; Johnson, Stewart, & Smith, 2012). Concussions are one, generally less severe, type of TBI, diagnosable within the three categories of mild, moderate, and severe (Spinal Cord, 2018). This form of TBI is of increasing presence with an estimated 79 concussed students at Rock Canyon High School in 2017 alone (G. Sims, personal communication, August 18, 2018). Overall, TBI is one of the leading causes of death and disability of youth in the United States, with 2.8 million diagnosed cases annually and an average mortality rate of 50,000 people per year (Taylor, *et al.*, 2017).

As human neurons regenerate slowly, the above hindrances can remain life long, severely altering and degrading one's quality of life. Currently no treatments are available to aid in the recovery of TBI. Consequently, there is a growing necessity to increase research on drug targets and the understanding of mechanisms of TBI. Therefore, model organisms are needed to initiate new studies. Research has been done on higher level mammalian organisms, such as mice and rats, regarding the neurological mechanisms of TBI (Cacialli, Palladino, & Lucini, 2018). Prior research performed on adult zebrafish, conducted on the neuro0logical effect of TBI, used a stab lesion assay demonstrating the ability of neuronal regeneration mechanisms, repairing damaged tissues and neurons in the impacted area (Schmidt, Beil, Strähle, & Rastegar, 2014). In other research, a weight dropped on the heads of adult zebrafish induced changes in the fish's movement, indicating TBI (Maheras, *et al.*, 2018). Although research previously used adult zebrafish, embryos are a potential candidate for future TBI research as model organisms for faster drug screening and observable developmental effects.

Zebrafish embryos are ideal model organisms because they are easy to care for and inexpensive to house in a research facility. One zebrafish breeding pair can produce up to 200 embryos per week (Zebra-fish Health, 2010). Accord-ing to guidelines from the National Institute of Health, zebrafish embryos are not considered vertebrates until 7 dpf (National Institute of Health, 2016). Up through 3 dpf, the embryos remain in a transparent chorion protecting them from the external environment **(Pic. 1)**. Due to their transparency, embryonic zebrafish development is well documented and understood **(Fig. 2)** (Your Genome, 2014). Approximately 70% of zebrafish genes are homologs to that of humans (Chitramuthu, 2013). Additionally, ze-brafish brains, which contain neurons similar to that of our own, have closely related develop-mental stages to humans.

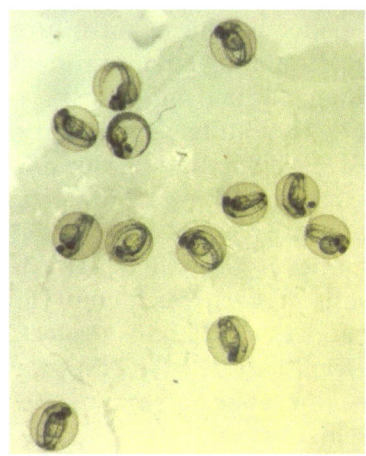

Picture 1: This image shows embryos before 3 dpf, as visibly seen within their clear chorions.

Zebrafish also have unique neuroregenerative properties that make them interesting subjects when looking at healing the wounds and damage associated with TBI. Scientists are striving to identify the genes and cellular processes involved with this neuroregeneration so that it can be applied to human treatment of neurological damage (Cacialli, *et al.*, 2018). However, there has been no research investigating zebrafish embryos ability to model TBI.

The purpose of this research was to identify if zebrafish embryos can receive TBI and thus serve as good model organisms in TBI research. It was hypothesized zebrafish embryos can be an effective model organism of TBI via treatment in a shaking incubator as used in this project. To induce TBI in zebrafish embryos, they were exposed to a shaking incubator method intended to cause head trauma and concussion-like symptoms. Measured next was the subsequent effects of mortality, neurological effects using a touch assay, and edema by measuring the head size within the treatment and control groups.

METHODS

This research analyzed the effectiveness of zebrafish embryos to serve as potential model organisms for TBI related research. To be an effective model, embryos had to demonstrate clear TBI symptoms following injury to the brain. In this research, the developing embryos were exposed to external forces designed to create neurological trauma to the brain. At the time of induction in the shaking incubator, all embryos had emerged from their chorion with brains capable of inducing damage (Hill, 2018). The TBI inducing

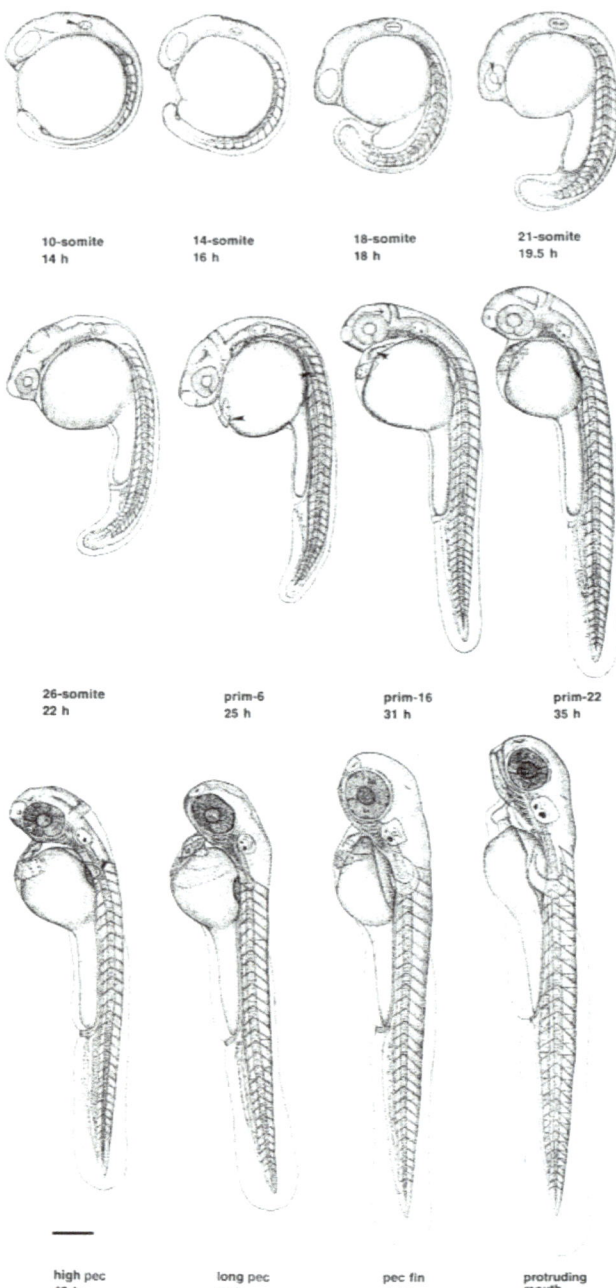

10-somite 14 h	14-somite 16 h	18-somite 18 h	21-somite 19.5 h
26-somite 22 h	prim-6 25 h	prim-16 31 h	prim-22 35 h
high pec 42 h	long pec 48 h	pec fin 60 h	protruding mouth 72 h

Figure 2: This is an illustration of zebrafish development. Reprinted from "Stages of Embryonic Development of the Zebrafish," by C. Kimmel, W. Ballard, S. Kimmel, B. Ullmann, & T. Schilling, 1995, *Development Dynamics: an official publication of the American Association of Anatomist, 203*(3), p. 258-259. Copyright 2005 by John Wiley and Sons. Reprinted with permission.

trials consisted of one TBI induced group and one control group of embryos that did not receive a TBI treatment. The subsequent mortality percent, reaction to stimuli, and head size were evaluated.

Zebrafish Care and Breeding

To properly care for the adult zebrafish, they were housed in two 10-gallon tanks; females were kept in one tank separate from the males in another **(Pic. 2)**. The water was at a constant temperature of 28.5 °C and between a pH of 6.8 to 7.5 in order to provide ideal conditions for breeding and

Picture 2: Female casper fish kept in a separate tank from zebrafish males at a constant temperature of 28.5 °C.

development. The pH as well as levels of ammonia, nitrate, and nitrite were measured biweekly using a colorimetric test. Ammonia and nitrite levels were as close to zero as possible, and nitrate was maintained at less than 50 mg/L. The water was cleaned every week and the filter checked to ensure proper function. The tanks were on a constant light cycle of 14 hours of light: 10 hours of darkness. Additionally, the tanks were in a low-traffic areas, as to minimize disruption. The adults were fed daily, with a diet consisting of fish flakes and dry brine shrimp; the protein was fed to the fish during attempts to breed. The night before breeding, one female and two males were placed into a breeding tank (Carolina Biological Supply Company), separated by a plastic divider, and left in complete darkness overnight (**Pic. 3**). The subsequent morning, the light turned on, and the divider was removed allowing the males and female access to breed. When the fish bred and eggs were visible, the adults were returned to their original tanks. The embryos were placed in a petri plate containing egg water within a hot water bath at 28.5°C inside a laminar flow hood, under natural lighting conditions. The egg water consisted of 1 L distilled water and 1.5 mL stock salts solution, composed of 40 g of Instant Ocean Aquarium Salts dissolved into 1 L of distilled water. (The Zebrafish Information Network, 2013).

Picture 3: The two adult male zebrafish (seen left), and the adult female casper fish (seen right) in the breeding tank separated by a barrier.

TBI Inducing Methods

To induce TBI, embryos were placed in a culture tube with 1 mL of egg water on the platform of a shaking incubator (**Pic**

4). The embryos were exposed to 250 rpm for 1 minute set to 28.5°C. In pretrials, it was determined 250 rpm caused the desired mortality of 10-25% similar to human TBI. Embryos in the control group underwent the same conditions but were not exposed to trauma in a shaking incubator. After induction, both groups were placed in separate wells of a 6 well petri plate with 5 mL of egg water again set in a hot water bath at 28.5°C until the assays were performed. All procedures, but TBI induction, were the same between the control and experimental group. To measure the effects of the TBI, three assays were performed.

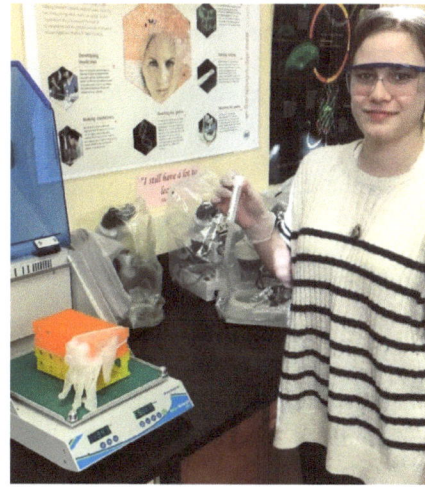

Picture 4: The embryos within a culture tube, being placed into the shaking incubator.

Mortality Assay

In order to measure the effects of the exposure to the TBI inducing event, a mortality as-say recorded the number of deceased embryos 24 hours post TBI as a percent of the total population of the experimental group. This number was compared to the non-trauma control. Once the method achieved the expected 10-25% mortality rate, it continued with the following touch and head size assays. Deceased embryos were differentiated from the living embryos through identification of a heart beat or lack thereof. To identify the lack or presence of a heart beat embryos were observed under a Leica KL300 LED dissecting microscope.

Touch Assay

Previous literature states touch assays assess muscle performance by analyzing the reaction to the stimulus. The touch assay helped conclude the development of the embryogenesis to evaluate the progress of the neurological growth (Sztal, Ruparelia, Williams, & Bryson-Richardson, 2016). Touch assays are a validated method to measure

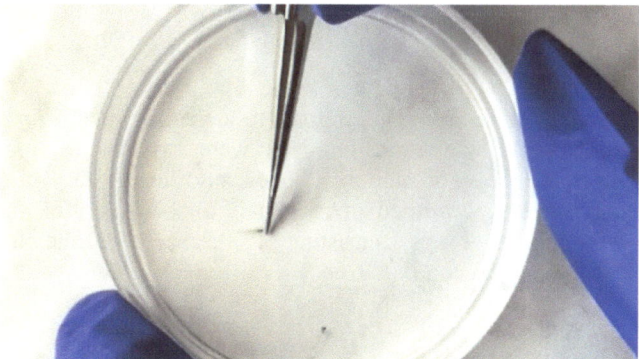

Picture 5: An example of the touch assay using the sharp tipped forceps to record reactions of the embryos.

neurological function in zebrafish and zebrafish embryos. To perform the touch assay, a sharp tipped forceps was used to lightly touch the embryo on its head twice (**Pic. 5**). The second touch was recorded, to eliminate the variable of shocked response, and rated on a scale of one to three where one = little to no reaction to the touch, two = a baseline or normalized reaction to the touch, and three = exaggerated and prolonged reaction to the touch. To avoid bias, the touch assays were rated blindly where two team members, who knew the treatment, performed and filmed the assay, and the other viewed the films at a later time and assigned the rating. This was done in order to eliminate, as the one rating an embryos reaction was not pre exposed to it.

Head Size Assay

The last assay performed measured the head size of the embryos at the fore, mid, and hindbrain, in proportion to the body length. In order to do this, the embryos were anesthetized 24 hours post treatment with single drops of tricaine administered to the egg water until the zebrafish embryos were immobilized. The embryos were imaged laterally using a LPl an 2X objective and the scale bar on the EVOS-FL Cell Imaging System to measure the average embryo head size and body length (**Pic. 6**). When comparing

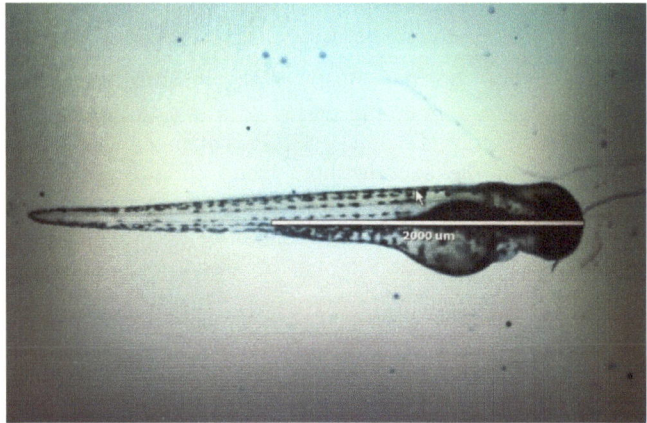

Picture 6: An embryo, at 2X objective on the EVOS-FL microscope, with scale bar aligned horizontally, used to measure the fore, mid, and hindbrain in proportion to the body length for the head size assay.

the experimental to control groups, signs of swelling would indicate a secondary injury associated with TBI previously shown in adult zebrafish (Cacialli, *et al.*, 2018). After they were imaged, embryos were transferred into a plastic bag and stored at - 20°C for a week before being disposed of as biological waste (Matthews & Varga, 2012).

RESULTS

After exposing *D. rerio* embryos to a shaking incubator to induce TBI symptoms, embryos were exposed to three different assays of mortality, touch, and head size. These assays were performed in two trials on both control and experimental groups, consisting of 30 embryos in the first and 22 in the second.

Mortality Assay

No embryos within the control group for either trial displayed mortality 4 dpf, corresponding to a 0% mortality rate. Within the first trial of the experimental group, 2 of the 15 embryos died, resulting in 13% mortality, corresponding within the 10-25% desired mortality. Of the 11 embryos in the second experimental group, another 2 died, producing 18% recorded mortality. Overall, both experimental groups for the mortality assay displayed within the 10-25% death allowing to continue the two assays for this inducing method.

Touch Assay

In this assay, the data from the first and second trials were combined to make the control and experimental group. Of the control group, 25 embryos were rated one, no reaction, only 1 embryo rated a two, and no embryo rated a three. The experimental group resulted in similar findings, recording 21 embryos with the rated reaction of one. Again, only 1 embryo recorded as having a reaction rated at two and 0 embryos a three (**Graph 1**).

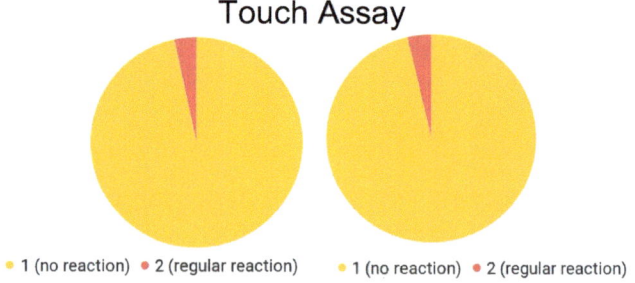

Touch Assay

- 1 (no reaction) ● 2 (regular reaction)   ● 1 (no reaction) ● 2 (regular reaction)

Graph 1: This graph shows the rated reactions of the embryos from the touch assay on a 1-3 scale from the experimental group (left) and control group (right).

Head Size Assay

Within the first trial, the control group had an overall average head to body length proportion of 0.442, while the experimental group had an overall proportion of 0.455. The second trial control group had an average head to body length proportion of 0.300 and the experimental group had an average overall proportion of 0.270 (**Graph 2**).

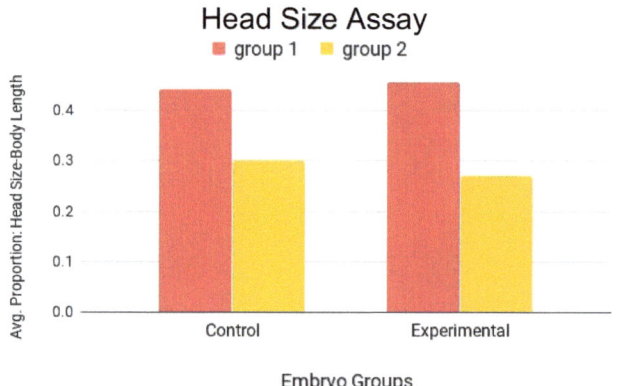

Head Size Assay
■ group 1   ■ group 2

Graph 2: The average head size of an embryo at the fore, mid, and hindbrain, proportionalized to the embryo's body length for groups 1 and 2.

DISCUSSION

This investigation explored zebrafish embryos potential as model organisms for future TBI research by identifying their ability to demonstrate TBI through assessing the secondary symptoms of death, reaction to a stimulus, and head edema.

It was hypothesized that zebrafish embryos could be effective model organisms of TBI when induced through treatment in a shaking incubator as used in this project.

Mortality was recorded 3 dpf between 10-25% for both experimental groups, the first trial producing 18% mortality and the second 13%, yet, 0% within both control groups. As there was the desired mortality of 10-25%, the embryos showed one possible symptom of TBI or at least the capability to sustain fatal injury. In the touch assay, both control and experimental groups had the majority of embryos rated reaction at a one while only a single embryo from each group was rated at a two and zero rated at a three for all groups. In the last assay, head size, the average proportion of the fore, mid, and hindbrain to the body length was 0.455 for the experimental and 0.442 for the control of the first trial. The second trial average head size to body length was 0.300 for the experimental and 0.270 for the control. However when looking at the response to stimulus and head size assay, no notable difference between the control and experimental groups was found. Therefore, within these two assays, there was no indication of TBI symptoms. Consequently, it can be surmised that the method used in this experiment of the shaking incubator at 250 rpm for 1 minute failed to produce TBI, disputing the hypothesis *D. rerio* embryos would show measurable TBI symptoms and thus the capability to serve as effective model organisms.

The lack of data could result from the shaking incubators inability to localize damage to the head, resulting in injury to the embryos physiology but not in the form of a TBI **(Pic. 7)**.

Moreover, the lack of localization in this method could lead to injury of the embryo in more than one area. Thus, an embryo with morphological damage could additionally sustain head damage in the form of TBI. As all morphological injured embryos were removed from a group during trials, any embryo with head injury, potentially TBI, in addition to other morphological damage along the body, would have been removed from the trials, thus skewing the data.

Picture 7: A 4 dpf embryo sustaining fatal damage to the spine resulting from treatment in the shaking incubator.

This research is applicable to the world and future research through contributing the knowledge that zebrafish embryos are capable of damage, in the sense of sustaining observable injury. If embryos are capable of damage then further research should be done to definitively identify if damage can be induced in the brain, thus resulting in a TBI. Errors within this research can be credited to limited experience in working with both adult zebrafish, embryos, and breeding techniques. Previous research working with zebrafish demonstrated that an embryos most important development

stages occurs within 3 dpf and that variation from the temperature of 28.5°C during these stages hinders development (Kimmel, Ballard, Kimmel, Ullmann, & Schilling, 1995). Temperature variation could have affected the embryos of this project as they were transferred from a breeding tank to petri plates in hot water baths within these fragile development stages. The temperature fluctuation could have led to the high amount of death seen within these stages, before induction, limiting the number of embryos per trial. Increasing research on zebrafish husbandry and care is needed to minimize variability that can occur within development and potential causes of death to embryos (Matthews & Varga, 2012). Maintaining consistent care concerning the embryos, such as sustaining temperature, could have helped regulate the amount of death seen in our research in the embryos before induction.

Another limitation of this research can be found in the lack of assays used to assess the potential presence of TBI in the embryos. One study monitored the effects of TBI in adult zebrafish through the use of a memory and swimming test apparatus. After TBI was induced, the spatial memories of the adult fish were assessed based on their ability to navigate the arena. This showed TBI induced fish had significant deficits in spatial memory (Maheras, *et al.*, 2018). Such behavioral assays were incapable to perform within this project as NIH guidelines required to euthanize all embryos before 7 dpf. These restrictions did not allot for the time needed to assess such behaviors. Other research done measured myelination in vivo through the use of a green fluorescent protein, allowing visualization of axons (Jung, *et al.*, 2010). Myelination loss, or damage, is indicative of TBI. Thus, the ability to image fluorescent myelination could have provided an in vivo identification of the presence of TBI. However, by breeding our own embryos without a genetic modification, we were unable to observe the fluorescent myelination under a microscope.

Future trials could follow the shaking incubator method, including and eventually optimizing different variables added to inflict TBI. Another step could include identifying and performing a method localizing and focusing damage to only the head of an embryo as to induce TBI. More embryos and time would have allotted for experimentation with different methods to induce TBI.

ACKNOWLEDGMENTS
I would like to thank Karlie Fedder for mentoring us through this project. I also want to thank Kayt Scott for providing valuable information such as how to set up our in home breeding tanks, general husbandry of both embryos and fish, and the fore. mid, and hindbrain locations. Further thanks goes to Vaishnavi Elango and Ashley Kozlowski for allowing me to collect data with them and be a member of their research team. Thanks goes to biotechnology teacher Shawndra Fordham who helped and overviewed my scientific process and overall project. I give my gratitude to Gwendolyn Karaba for aiding in determining how to collect and analyze data. Kerry Hinton, the Chemical Safety Manager, is another person I would like to thank for making sure the chemicals used in this project were safe to work with

in a BSL-1 lab. I would also like to thank Bryan Winkelman, RCHS Teacher Librarian, for assistance with setting up a website for my research in addition to all the other help he has given me. I would like to show appreciation to Kristin Karnicki for helping in the overall betterment of this project. I thank Peter Thompson for providing insight on how to improve this project. Additionally, I want to thank Nayan Naik and the following families for their generous donations that without our project would not have been possible: Kozlowski's, Elango's, and Zilligen's. I would like to express appreciation to Rock Canyon High School and Douglas County School District for providing space and equipment to perform my research.

REFERENCES

Andrews, M. (2012). Concussion Anatomy.png. https://commons.wiki media.org/wiki/File:Concussion_Anatomy.png

Cacialli, P., Palladino, A., & Lucini, C. (2018). Role of brain-derived neurotrophic factor during the regenerative response after traumatic brain injury on adult zebrafish. *Neural Regeneration Research, 13*(6), 941-944. doi:10.4103/1673-5374.233430

Chitramuthu, B. (2013). Modeling Human Disease and Development in Zebrafish. *Human Genetics & Embryology, 3(*1). doi:10.4172/2161-0436.1000e108

Hill, M. (2018). Embryology. *Zebrafish Development.* Retrieved from https://embryology.med.unsw.edu.au/embryology/index.php/Zebra fish_Development

Johnson, V., Stewart, W., & Smith, D. (2013). Axonal Pathology in Traumatic Brain Injury. *Experimental Neurology, 246,* 35-43. doi:10.1016/j.expneurol.2012.01.013

Jung, S. , Kim, S. , Chung, A. , Kim, H. , So, J. , Ryu, J., ...Kim, C. (2010). Visualization of myelination in GFP-transgenic zebrafish. *Developmental Dynamics, 239*(2), 592-597. doi:10.1002/dvdv.22166

Katz, D., Cohen, S., & Alexander, M. (2015). Mild Traumatic Brain Injury. *Handbook of Clinical Neurology, 127*(9), 131-156. doi:10.1016/B978 -0-444-52892-6. 00009 -X

Kimmel, C., Ballard, W., Kimmel, S., Ullmann, B., & Schilling, T. (1995). Stages of Embryonic Development of the Zebrafish. *Developmental Dynamics, 203*(3). 253 -310. doi:10.1002/aja.1002030302

Matthews, M. & Varga, Z. (2012). Anesthesia and Euthanasia in Zebrafish. *Journal of the Institute for Laboratory Animal Research, 53*(2).192–204. doi:10.1093/ilar. 53.2.192

Maheras, A., Dix, B., Carmo, O., Young, A., Gill, V., Sun, J., … Spence, R. (2018). Genetic Pathways of Neuroregeneration in a Novel Mild Traumatic Brain Injury Model in Adult Zebrafish. *eNeuro, 5*(1). doi:10.1523/eneuro.0208-17.2017

Murthy, T., Bhatia, P., Sandhu, J., Prabhakar, T., & Gogna, R. (2005). Secondary Brain Injury: Prevention and Intensive Care Management. *Indian Journal of Neurotrauma, 2*(1), 7-12. doi:10.1016/S09730508 (05)80004-8

National Institute of Health. (2016) Guidelines for Use of Zebrafish in the NIH Intramural Research Program, NIH, Intramural Research Program. (2016). [PDF file] Retrieved from https://oacu.oir.nih.gov/ sites/default/files/uploads/arac-guidelines/zebrafish.pdf

Prins, M., Greco, T., Alexander, D., & Giza, C. (2013). The pathophysiology of traumatic brain injury at a glance. *Disease Models & Mechanisms, 6*(6). 1307-1315. doi: 10.1242/dmm.011585

Rudy, S., Maia, P., & Kutz, J. (2016). Cognitive and behavioral deficits arising from neurodegeneration and traumatic brain injury: a model for the underlying role of focal axonal swellings in neuronal networks with plasticity. *Journal of Systems and Integrative Neuroscience, 2*(1), 114-121. doi:10.15761/JSIN.1000120

Schmidt, R., Beil, T., Strähle, U., & Rastegar, S. (2014). Stab wound injury of the zebrafish adult telencephalon: a method to investigate vertebrate brain neurogenesis and regeneration. *Journal of Visualized Experiments : JoVE, 1*(90)*,* e51753. doi:10.3791/51753

Spinal Cord. (2018). Types of Traumatic Brain Injury. Retrieved from https://www.spinalcord.com/types-of-traumatic-brain-injury

Sztal, T., Ruparelia, A., Williams, C., & Bryson-Richardson, R. (2016). Using Touch-evoked Response and Locomotion Assays to Assess Muscle Performance and Function in Zebrafish. *Journal of Visualized Experiments,* (116). doi:10.3791/54431

Taylor, C., Bell, J., Breiding, M., & Xu, L. (2017). Traumatic Brain Injury-Related Emergency Department Visits, Hospitalizations, and Deaths - United States, 2007 and 2013. *Surveillance Summaries, 66*(9), 1-16. doi:10.15585/mmwr.ss6609a1

The Zebrafish Information Network. (2013). General Methods for Zebrafish Care. Retrieved from https://zfin.org/zf_info/zfbook/ chapt1/1.3.html

Your Genome. (2014).Why use the zebrafish in research? Retrieved from https://www.yourgenome.org/facts/why-use-the-zebrafish-in-research

Zebrafish Health. (2010). Zebrafish Facts. Retrieved from http://www.zf-health.org/information/factsheet.html

ABOUT THE AUTHOR

Pictured: The author of this article, Zilligen is seen left, and Fedder, the mentor of this project, is seen right.

Zilligen found this research on identifying zebrafish embryos' ability to sustain TBI extremely pertinent as someone invested in scientific discovery and rugby, a TBI-producing activity. Zilligen sees her teammates hindered by TBI in the form of concussions, contributing to her passion for conducting TBI research. Zilligen enjoyed the experience, yet faced many challenging setbacks along the way. Foremost of these setbacks was the organisms with which she worked. Many times the zebrafish were unwilling to cooperate, setting research awry. This taught Zilligen problem solving and perseverance, skills she will appreciate in her near future as she participates in research projects in college and eventually, as a career. This project taught Zilligen that science is not about how much you know, but about how much you want to know.

5810 McArthur Ranch Road
Highlands Ranch, CO 80124
303-387-3000

Principal
Andy Abner
Andrew.Abner@dcsdk12.org

Registrar
Susan Delgado
Susan.Delgado@dcsdk12.org

Administrative Assistant
Barb Cocetti
Barbara.Cocetti@dcsdk12.org

STEM PROGRAMMING

The Principals of Experimental Design in Biotechnology course is one of many courses offered as part of our choice-driven STEM programming, which allows each of our students to prepare for their vision of a career in science, technology, engineering, or math.

Due to the competitive nature of STEM majors in college, we believe that taking a rigorous course load, including Honors, AP, and dual credit courses, is the best way to prepare students for the coursework they will encounter. In addition, involvement in clubs that encourage competition in Science, Technology, Engineering, and Business allows students the opportunities to think on their feet, construct and communicate arguments, and work through the engineering process. Finally, our wish is that students will become involved in an internship or shadowing experiences in order to gain the workplace experience that our classes may not provide.

Students who diligently pursue this difficult course load, as well as meeting these additional requirements, will not only benefit from their knowledge and preparation, but will also be able to show the universities that they are determined students by presenting them with a STEM certificate.

Canyon High School
Home of the Jaguars

Our Mission:
To Empower, To Explore, To Encourage and To Excel in Education

Our Vision:
Our student-centered culture practices collaborative decision making and continuous improvement in a safe, supportive environment.

Rock Canyon is a comprehensive high school consisting of grades nine through twelve, located in Highlands Ranch, Colorado, a southern suburb of Denver. Our community is composed primarily of working professionals.

Rock Canyon is part of the Douglas County School District (DCSD), the third largest school district in Colorado, serving over **68,000** students for the 2018-2019 school year. The district is comprised of 9 high schools, 9 middle schools, 47 elementary schools, 12 charter schools, 2 magnet schools, 3 alternative schools and an online school. The DCSD continues to maintain its standing as one of the finest, highest achieving districts in Colorado.

Rock Canyon opened in 2003. It has a current enrollment of 2,280 students. RCHS occupies a 279,250 square foot building on an 80-acre campus.

Rock Canyon High School prides itself on excelling in academics, activities and athletics to create a balanced and comprehensive high school experience for all students. We strive to develop a tradition of excellence in order to develop a premier high school program for all post-secondary options. Rock Canyon is currently ranked as one of the top high schools in Colorado.

We invite our parents to take an active role in their student's education by empowering their students to explore the many opportunities offered at Rock Canyon while continuing to encourage their students to excel in their educational goals. We truly believe a partnership must exist between the school and the family; together we can elevate our students to the next level.